Science Unlimited?

Science Unlimited?

The Challenges of Scientism

Edited by
Maarten Boudry
and Massimo Pigliucci

The University of Chicago Press
Chicago and London

The University of Chicago Press, Chicago 60637
The University of Chicago Press, Ltd., London
© 2017 by The University of Chicago
Published 2017
Printed in the United States of America

26 25 24 23 22 21 20 19 18 17   1 2 3 4 5

ISBN-13: 978-0-226-49800-3 (cloth)
ISBN-13: 978-0-226-49814-0 (paper)
ISBN-13: 978-0-226-49828-7 (e-book)
DOI: 10.7208/chicago/9780226498287.001.0001

Library of Congress Cataloging-in-Publication Data

Names: Boudry, Maarten, 1984– editor. | Pigliucci, Massimo, 1964– editor.
Title: Science unlimited? : the challenges of scientism / edited by Maarten Boudry and
    Massimo Pigliucci.
Description: Chicago : The University of Chicago Press, 2017. | Includes index.
Identifiers: LCCN 2017017535 | ISBN 9780226498003 (cloth : alk. paper) | ISBN
    9780226498140 (pbk. : alk. paper) | ISBN 9780226498287 (e-book)
Subjects: LCSH: Scientism. | Science—Philosophy. | Pseudoscience.
Classification: LCC Q175 .S36235 2017 | DDC 501—dc23
    LC record available at https://lccn.loc.gov/2017017535

♾ This paper meets the requirements of ANSI/NISO Z39.48–1992 (Permanence of Paper).

CONTENTS

INTRODUCTION

More than a century years ago, C. S. Peirce wrote that "a man must be downright crazy to deny that science had made many true discoveries" ([1903] 1997, 230). Nowadays, even more so than then, science is widely recognized as one of the marvels of the human intellect. It is our most reliable source of knowledge about the world. But what, if any, are the limits of science? Are there questions that will forever elude our best scientific efforts? And what is this thing we call science, anyway?

In discussions about the proper place of science, one often hears about an alleged sin called scientism. Science is a good thing, so the argument goes, but even good things can be pushed too far, beyond their proper limits. Can science settle *all* interesting questions? Should we leave room for other ways of knowing and understanding besides science, for example the methods employed in the humanities or in philosophy, or does science rule supreme in all these domains?

The limits of science is a topic of public interest, as witnessed by a spate of articles in popular magazines and blogs, such as the widely read discussion between Steven Pinker (2013) and Leon Wieseltier (2013) in the *New Republic,* op-eds by Ross Douthat in the *New York Times* and Oliver Burkman in the *Guardian,* or popular books by Alex Rosenberg (2011), Sam Harris (2011), and

Susan Haack (2007). More academic discussions of scientism include book-length treatments by Tom Sorell (2002) and Joseph Margolis (2003), and a recent volume criticizing scientism as the "new orthodoxy" (Williams and Robinson 2014).

All these authors use the concept of scientism in one way or another, but there is little agreement on its definition. At a first approximation, current usage suggests the following characteristics: (1) an excessive deference toward the deliverances of science and anything to which the honorific label *scientific* is attached; (2) brazen confidence in the future successes of scientific investigation, for example in arriving at a Theory of Everything, or in solving every interesting question about reality; (3) the conviction that the methods of science are the only worthwhile modes of inquiry, and will eventually supplant all others; (4) the thesis that other disciplines should be either subsumed under science or rejected as worthless; (5) the thesis that all ways of acquiring knowledge and understanding are (or should be) scientific in nature, and hence there is no interesting difference between science and other forms of inquiry.

Given the ubiquitous usage of the term in a wide variety of contexts, we think a frank philosophical discussion about the limits and proper place of science is long overdue. This volume explores the relationship between science and other ways of knowing, and the possible value of a concept like scientism to describe various forms of (excessive) science enthusiasm. It provides a forum for philosophers and scientists, both detractors and enthusiasts of Science Unlimited, to talk about the nature and scope of science.

Demarcation Problems

The question of the demarcation between science and nonscience has fascinated philosophers ever since Karl Popper ([1959] 2002) proposed the criterion of falsifiability as the distinguishing hallmark of good science. As it is traditionally understood, the demarcation problem concerns the borders between science and pseudoscience—that is, theories that pretend to be scientific, but do not live up to the standards of good science. We have put together a volume on this classical demarcation project, breathing new life into an old chestnut that had been prematurely pronounced dead by some philosophers of science (Pigliucci and Boudry 2013).

But there is another, related demarcation problem shimmering through in Popper's writing on the subject, even though he did not always keep the two demarcation jobs apart (Boudry 2013). This is the conceptual separation between science and other epistemic endeavors that are valuable in their own right. It concerns neighboring academic disciplines, such as history, philosophy, and mathematics, but also metaphysics and everyday knowledge acquisition. This volume deals with this second demarcation project: not the one that tries to distinguish science from its phony contenders, but the one of figuring out the limits of science itself. In this sense, the pages that follow provide a suitable companion to our earlier collection, *Philosophy of Pseudoscience*.

There is another way to spell out the link between the two demarcation problems in philosophy of science. Pseudoscience is a form of fool's gold: it looks like the real thing, but on closer inspection it turns out to be worthless. The problem with pseudosciences like astrology, homeopathy, and creationism is that they are *lacking* in scientific rigor; they fail to gain the evidential support we demand from science. The problem with scientism, by contrast, can be seen as an *excess* of science. To be guilty of scientism is to be overly deferential toward science, to unfairly disparage other disciplines like the humanities or philosophy, or to have an inordinate confidence in the future successes of science.

Still, how much of science is *too* much? And how far exactly can science venture before it *over*reaches? In discussions about its proper purview, a variety of authors have pronounced several domains and types of questions off-limits to science: ethics, subjective experience, abstract entities, intuitive knowledge, qualia, transcendence, metaphysics, and of course God. If it is true that science cannot penetrate those realms, does that mean that there are other ways of knowing, perhaps on an equal par with science, that can proffer insight into them?

As should be clear by now, the common usage of *scientism* is derogatory. Though originally the term simply denoted a scientific habit of thinking, or the worldview of a scientist, over time its meaning has shifted, acquiring a negative connotation, much like the evolution of the words *fundamentalism* and *reductionsim*. Nowadays *scientism* is most commonly used as a term of abuse in the same way as words such as *pseudoscience* and *superstition*. As with all normative concepts, however, it is itself susceptible to abuse. Those who stand accused of scientism typically dismiss the accusation as vacuous or meaningless, or as a disguised expression of antiscience. Indeed, even those who believe that *scientism* is a useful term to navigate our modern intellectual landscape admit that it has been eagerly exploited in the service of antiscience or pseudoscience. Hurling around the term *scientism* is sometimes little more than a convenient way to shield questionable doctrines and practices from critical scrutiny. It is hardly surprising, therefore, that many quack therapists, spiritualists, postmodern science critics, new age gurus, and theologians have pressed the term into service for exactly such purposes (Sheldrake 2012; Mc-Grath and McGrath 2007). Indeed, some argue that the usage of *scientism* as a cloak for antiscience is now so pervasive that the term can no longer be salvaged.

But there is an interesting twist to this story of the descent of *scientism* into a term of abuse. In an act of defiance, some science enthusiasts—both philosophers and scientists—have recently embraced the label as a badge of honor (Ladyman and Ross 2007; Rosenberg 2011). These proud offenders are calling on philosophers and humanities scholars to pay proper deference to the superior epistemic methods of the sciences. Science, they argue, should be the model of all human knowledge. In a way, these proponents of scientism are reclaiming something like the original and positive meaning of the term, as denoting a superior way of thinking and reasoning. Some have explicitly stated that all interesting questions can either be answered by science or not answered at all, and that the methods of (natural) science should prevail everywhere. Science is the only game in town, and there are no "other ways of knowing" worthy of that name. At best, there can be temporary placeholders for scientific knowledge, paving the ground for scientific achievements. Some self-proclaimed advocates of scientism argue that science will eventually reign supreme and supplant all other endeavors, while others affirm that it has *already* done so.

A related, but more moderate position claims that science is continuous with other forms of knowing, with no crisp borders between it and, for example, philosophy or the humanities or everyday reasoning. Scientific methods should be applied wherever possible, and territorial disputes and border patrols should be dismissed. Critics of scientism, however, tend to see this as a yet another form of scientific imperialism. For them, downplaying the differences between science and other ways of knowing, or dissolving the boundaries between them, is just another way to make science engulf all other disciplines, without any regard for differences in methodology and ways of understanding.

Even though the usage of *scientism* as a derogatory term is relatively recent, worries about the continuing progress of science are hardly new. Every scientific advance has been hailed by some as much as it has been loathed by others. Where some see an exciting foray into unknown territory, others see encroachment on already occupied terrain (Pinker 2013; Wieseltier 2013). For example, theology has retreated whenever science has become capable of probing the parts of nature where a divine presence was previously assumed. More controversially, some have argued that philosophers, too, have relinquished much of their territory to the sciences, with ethics turning into moral psychology, epistemology into cognitive science, and metaphysics into cosmology. Another way to tell the same story, of course, is to say that science itself started out as a branch of natural philosophy. "Science" is what we call philosophy when it matures into relatively independent disciplines, with specialized methods, procedures, and institutional arrangements. Perhaps philosophers should applaud this as progress, and look forward to the day when their discipline evaporates completely, blossoming into so many sciences. Others would reject this line of reasoning as yet another instance of scientism, which fails to do justice to the proper nature of philosophy (Pigliucci 2016).

Many critics of scientism worry that the public prestige and authority of science in modern society will eventually drive the humanities and philosophy to extinction. Funding budgets are already shrinking, and politicians' increasing demands for quantifiable results and technological applications from academic teaching and research are putting further pressure on what used to be called a liberal arts education. Consequently, discussions of scientism are not just academic but very much relevant to the education, and ultimately the lives, of millions of students worldwide.

Two Approaches to the Limits of Science

This volume aims at exploring the limits and nature of science, the territorial disputes arising from scientific progress, and the proper stature of science in modern culture. Is science endangering other forms of inquiry? Is the term *scientism* a useful guide to navigate our current intellectual landscape? Does it capture an interesting phenomenon, and if so, is it something to worry about or to rejoice in? How should we distinguish between legitimate uses of the term *scientism* and self-protective and unsubstantiated ones? Is the preoccupation with scientism itself entirely misguided?

In this collection, two broad approaches to the question of scientism and the limits of science emerge. The first strategy starts from a normative position, treating scientism as an intellectual sin. That we should disapprove of scientism is then a trivial truth, taken as a point of definition (Haack 2012). *Scientism* just means too much science, or science overreaching. With this normative definition of *scientism* in place, it remains of course a substantive and interesting question whether there is anything in our modern culture that fits the description of scientism. There is a range of viewpoints within this approach. At one end of the spectrum, some contributors argue that science clearly has limits, and that scientism is a genuine cause for concern, or even a major threat to the intellectual health of modern society. These authors want to protect various domains or disciplines from what they consider to be scientific imperialism. Such domains include philosophy (Pigliucci, Sorell), folk psychology (Buekens), and metaphysics or religion (Ruse). Others are arguing not for principled limits to science but merely for practical and historical limits. Scientism then consists in the neglect of those practical limits (see Nickles's chapter on strong realism). Some of our authors argue that there may be no identifiable limits to science, but there are other intellectual purposes and ways of life besides science. Appropriating morality to science, or expecting salvation from science in the domain of politics, may then be seen as forms of scientism (Edis). Finally, some contributors start from the pejorative definition of *scientism*, meaning illicit attempts to overstep the limits of science, but then conclude that there are no interesting limits to science, so that the question of scientism evaporates (Boudry, to some extent Ross).

The second general approach, rather than taking a normative starting point, tries to give a neutral description of scientism as an ordinary philosophical position akin to naturalism, dualism, transcendentalism, physicalism, and so on. From this point of view, it is an open question whether the tenets of scientism, suitably defined, are defensible or not. It is here that we find those philosophers sporting the term *scientism* as a badge of honor (Rosenberg, Ross). In their eyes, scientism is not a sin to be avoided but a position to be outlined and defended. Still other authors start from a neutral definition of *scientism*, but then level arguments against it (Peels, Kalef, to some extent Buekens). For them, scientism is to be rejected, but not as a matter of stipulative definition.

In short, we can distinguish between two stances. Either one starts off by treating *scientism* as a term of abuse, and then proceeds to see if anything fits the bill; or one starts by defining *scientism* as an ordinary philosophical position concerning the limits of science, after which one examines whether or not it is defensible. It should be noted that these two strategies partly overlap, and do not preclude substantial agreement between authors on what scientism roughly amounts to. Cutting across the normative and the nonnormative approaches, we see authors referring to scientism as the notion that science, in one way or another, is the best or only game in town. Different variations on that thesis include (1) if a question can be answered at all, it can be answered by science; (2) there are no valid modes of inquiry besides science; (3) all human knowledge is ultimately scientific. For some contributors to the volume, such notions are interesting theses to be explored, and to be argued for and against. For others, they are obvious forms of hubris that overstep the proper boundaries of science, and do no favors to science itself.

Finally, it should be noted that several authors in this volume have misgivings about the term *scientism* itself. Some do so because it is often used as a cover for antiscience (Law, Edis), others because it presupposes identifiable limits to science that simply do not exist (Boudry), still others because it is too generic and vague as a diagnostic label (Blackford), or all these reasons combined. These authors agree that too much of anything is bad, including too much science, but do we need a label like *scientism*? Enterprises that falsely adopt the trappings of science are already known under a more familiar name: pseudoscience. And for the excessive deference to (natural) science, more specific diagnostic labels could be proposed, such as *mathematicism* (Blackford), *neurohyping*, or *philosophy bashing* (Pigliucci).

However, even if we ultimately reject the term *scientism* for one or another reason, we still need to think about the nature and the limits of science. No contributor to this volume disputes that science has by and large been a spectacularly successful endeavor, and that it is one of the crowning achievements of the human intellect. Precisely because it is so powerful, however, science can also be dangerous. And precisely because science has historically extended its reach ever farther, it is tempting to conclude it knows neither bounds nor limits. Whether this thesis is on the right track or dangerously mistaken remains to be seen.

Either way, the hope of the editors and contributors to this volume is to advance the discussion of scientism and the nature of science by way of a genuine philosophical debate, reflecting the panoply of intellectual positions as well as providing guidance to academics, policy makers, and the public at large about the nature of the debate and its many implications.

One final note by way of guide to the reader: the chapters are ordered alphabetically by their author's surname, as there is no particular order of exposition. It is possible to read the book from cover to cover, or to skip back and forth between chapters. A number of chapters may require some technical background in philosophy of language (Buekens), economics (Ross), and epistemology (Sorell), but most are accessible to a lay audience without prior knowledge in philosophy or epistemology. We hope this volume will stimulate further discussion on scientism and the limits of science, both inside and outside academic circles.

Maarten Boudry and Massimo Pigliucci
Ghent and New York City, January 2017

······

Literature Cited

Boudry, Maarten. 2013. "Loki's Wager and Laudan's Error: On Genuine and Territorial Demarcation." In *Philosophy of Pseudoscience: Reconsidering the Demarcation Problem*, edited by M. Pigliucci and M. Boudry, 79–98. Chicago: University of Chicago Press.

Haack, Susan. 2007. *Defending Science—within Reason: Between Scientism and Cynicism*. Amherst, NY: Prometheus Books.

———. 2012. "Six Signs of Scientism." *Logos and Episteme* 3 (1): 75–95.

Harris, Sam. 2011. *The Moral Landscape: How Science Can Determine Human Values*. New York: Simon and Schuster.

Ladyman, James, and Don Ross. 2007. *Every Thing Must Go: Metaphysics Naturalized*. Oxford: Oxford University Press.

Margolis, J. 2003. *The Unraveling of Scientism: American Philosophy at the End of the Twentieth Century*. Ithaca, NY: Cornell University Press.

McGrath, Alister, and Joanna Collicutt McGrath. 2007. *The Dawkins Delusion? Atheist Fundamentalism and the Denial of the Divine*. Downers Grove, IL: InterVarsity Press.

Peirce, C. S. (1903) 1997. *Pragmatism as a Principle and Method of Right Thinking: The 1903 Harvard Lectures on Pragmatism*. Albany: State University of New York Press. Citations are to the 1997 edition.

Pigliucci, Massimo. 2016. "The Nature of Philosophy: The Full Shebang," May 30. *Plato's Footnote* (blog), https://platofootnote.wordpress.com/category/nature-of-philosophy/.

Pigliucci, Massimo, and Maarten Boudry, eds. 2013. *Philosophy of Pseudoscience: Reconsidering the Demarcation Problem*. Chicago: University of Chicago Press.

Pinker, Steven. 2013. "Science Is Not Your Enemy." *New Republic*, August 6, https://newrepublic.com/article/114127/science-not-enemy-humanities.

Popper, K. R. (1959) 2002. *The Logic of Scientific Discovery*. London: Routledge.

Rosenberg, Alex. 2011. *The Atheist's Guide to Reality: Enjoying Life without Illusions*. New York: W. W. Norton.

Sheldrake, Rupert. 2012. *The Science Delusion*. London: Coronet.

Sorell, Tom. 2002. *Scientism: Philosophy and the Infatuation with Science*. London: Routledge.

Wieseltier, Leon. 2013. "Crimes against Humanities: Now Science Wants to Invade the Liberal Arts; Don't Let It Happen"; response to Pinker (2013). *New Republic*, September 3, https://newrepublic.com/article/114548/leon-wieseltier-responds-steven -pinkers-scientism.

Williams, Richard N., and Daniel N. Robinson, eds. 2014. *Scientism: The New Orthodoxy*. London: Bloomsbury Academic.

1

The Sciences and Humanities in a Unity of Knowledge

RUSSELL BLACKFORD

Scientism Talk

Much of the concern about a sinister intellectual tendency called scientism involves the relationship between the sciences and the humanities. Scholars in disciplines such as history, philosophy, and literary studies appear nervous about loss of political support and erosion of public funding. Their nervousness is justified, because there's a widespread cynicism about the humanities, some of which may indeed result from glorification of science and technology.

The philosopher of science John Dupré complains that much philosophical work in metaphysics shows an unhealthy reverence for science. He suggests that *science*—when "construed broadly and preanalytically"—"means little more than whatever are our currently most successful, or even just influential, ways of finding out about particular ranges of phenomena." He adds, however, that "abuse of an excessively rigorous and restrictive conception of science is part of what I mean by the (intentionally abusive) term *scientism*" (Dupré 1993, 167). As we'll see, the popular and historical meaning of the word *science* is indeed relatively narrow and "restrictive." Confusion—or certainly the "abuse" that Dupré refers to—arises only if it's thought that all legitimate inquiry falls into that restricted area.

Dupré writes facetiously of "the unity of scientism" to refer to the sociological unity of people who are institutionally certified as scientists. This tends, he claims, to give certified scientists and their work a dubious epistemic authority (1993, 222). He adds that another aspect of scientism could be called mathematicism (223): the enhanced prestige accorded to those parts of science that give mathematical methods a central role. (As an alternative, we could call this tendency "mathematics fetishism.")

Dupré's views follow, in part, from his rejection of the idea of an ultimately orderly universe. Yet we need not adopt his specific approach to metaphysics to suspect that he has a point about mathematics. Some fields of inquiry produce reliable results with little in the way of mathematical systematization (though they may demand skilled and arduous detective work). These fields may be unfairly denied academic and popular prestige. Other fields—perhaps some areas of the social sciences—may be dauntingly mathematical, yet relatively unimpressive in their empirical results. Without mathematicism, Dupré claims, "substantial parts, at least, of a number of disciplines would sink without a trace" (1993, 224).

Much science is of course precise in its methods and robust in its core findings. Modern science's use of sophisticated mathematics has been integral to its success. But Dupré's concerns about current intellectual trends do not appear merely silly or specious. Among these trends, we often see an unhelpful mimicry by the social sciences and humanities of the superficial trappings of the natural sciences. At the same time, strangely or not, we can see widespread resentment of science (within some humanities departments, of course, but also in the culture at large). Following Susan Haack (2007, 18–19), we can identify various kinds of science envy and antiscience—with the latter manifesting as hostility to science or as science denialism.

Indeed, *pace* Dupré, we might wonder whether institutionally certified scientists, or at least some of them, should actually be accorded *more* epistemic authority by the general population. Consider the political successes of some kinds of science denialism, especially denial of climate change. Before we worry too much about glorification of technoscience, is it really a *good* thing that inconvenient scientific findings can be impugned so easily and effectively in current political debates? Conversely, would it have been a *bad* thing if the current generations of voters in Western liberal democracies had been socialized to have more respect for scientific consensus?

Among all this, much scientism talk—weaponized accusations of scientism—comes from theologians and religious apologists. These individuals do not merely strive to defend the humanities. They contend, rather, for supernatural or nonrational ways to obtain knowledge about the world and the human condition. John F. Haught is just one of many theologians who accuse Richard Dawkins and other publicly outspoken atheists of scientism (e.g. 2008, 18–19, 63). Haught views scientism as a belief that scientific methodology—whatever exactly that might be—can answer all questions, including those relating to meaning, values, and the existence of God. Another high-profile theologian and fan of the word *scientism*, Alister McGrath, explicitly defines it along similar lines: "The clumsy word 'scientism'—often glossed as 'scientific imperialism'—is now used to refer to the view that science can solve all our problems, explain human nature or tell us what's morally good" (2011, 78).

In the following sections, I argue that we can defend humanistic scholarship without endorsing any mysterious "ways of knowing." We needn't, for example, believe that knowledge can be obtained through divine revelation, recourse to holy books, mystical rapture, or faith (defined in a manner that contrasts with reason),[1] or via a built-in *sensus divinitatis* that gives us an immediate apprehension of God.

I propose that we abandon the word *scientism*. In the next section, I'll start with a closer look at this difficult word.

Of Scientism, Science, and Scientists

As philosophers are invariably (and painfully) aware, dictionaries have their limitations. To say the least, they can be unhelpful for pinning down the nuances of concepts. There is, perhaps, something old-fashioned and out of favor about recourse by philosophers to dictionary definitions (see Sorell, this volume). On this occasion, however, I am heavily indebted to the *Oxford English Dictionary*. It provides plausible and illuminating definitions of *scientism* and related words such as *science* and *scientist*. In fact, the *OED* offers two definitions of *scientism*. For the record, the first: "A mode of thought which considers things from a scientific viewpoint." (Well, that sounds harmless enough!) For present purposes, however, we should focus on the word's complex second definition:

> Chiefly *depreciative*. The belief that only knowledge obtained from scientific research is valid, and that notions or beliefs deriving from other sources, such as religion, should be discounted; extreme or excessive faith in science

or scientists. Also: the view that the methodology used in the natural and physical sciences can be applied to other disciplines, such as philosophy and the social sciences.

The *OED* traces this usage back to 1871. Its definition is helpful in contrasting scientific research with other putative sources of knowledge such as, specifically, religion. As we'll see, however, the definition raises further issues. I'll soon return to them, but for a start, what counts as specifically *scientific* research, and what, exactly, is the "methodology used in the natural and physical sciences"?[2]

Meanwhile, the *OED* provides several definitions of *science* itself, allowing some room for Dupré's broad, preanalytical construal. However, the dictionary indicates that since the mid-nineteenth century, the most usual meaning of *science* when the word is employed without qualification is narrower:

> The intellectual and practical activity encompassing those branches of study that relate to the phenomena of the physical universe and their laws, sometimes with implied exclusion of pure mathematics. Also: this as a subject of study or examination.

This usage dates back to 1779. Notably, it is the definition for which the *OED* asks us to compare the depreciative usage of *scientism*.

Also relevant, perhaps, is the concept of *a science*, for which the *OED* defines a relevant usage that it traces to 1600:

> A branch of study that deals with a connected body of demonstrated truths or with observed facts systematically classified and more or less comprehended by general laws, and incorporating trustworthy methods (now esp. those involving the scientific method and which incorporate falsifiable hypotheses) for the discovery of new truth in its own domain.

As for the word *scientist*, see the following: "A person who conducts scientific research or investigation; an expert in or student of science, esp. one or more of the natural or physical sciences." According to the *OED*, this usage dates to 1834 and 1840: to proposals first made in print by William Whewell, who sought a word that could be used to "designate the students of the knowledge of the material world collectively." He wanted something narrower than *philosopher*, but broader than the words referring to practitioners of specific sciences. Thus, *scientist* was coined by analogy to *artist*. In Whewell's understanding of the latter term, it included musicians, painters, and poets, among others. Whewell considered scientists to include, as examples, mathematicians, physicists, and naturalists.[3]

This cluster of definitions from the *OED* captures the popular and historical understandings of the terms *science* and *scientist* without the need for endless introspection and conceptual analysis. For example, it makes sense of the common distinction between a university's faculty of science and its other faculties (often including a faculty of "arts" or "liberal arts" or "humanities"—with the social sciences frequently being housed with the humanities rather than within the science faculty).

As already noted, the *OED* also captures the theological element in much nineteenth-, twentieth-, and twenty-first-century complaint about scientism. More generally, it includes the oft-expressed suspicion that there is an excessive deference to science. Still, dictionary definitions do not tell us what kind of "faith in science or scientists" counts as "extreme or excessive." The *OED* also leaves unclear whether or not mathematics is part of science ("sometimes with implied exclusion of pure mathematics"), and whether or not mathematicians are scientists (Whewell evidently thought they were). It is unclear from the *OED* definitions whether individuals who make accusations of scientism imagine its proponents to include mathematical theorizing among their "valid" sources of knowledge.

If it comes to that, how should we regard historical scholarship, much of which consists of locating, translating, and reconciling inscriptions and documentary records? This is not usually regarded as a form of scientific research. But do the scientism accusers imagine that their antagonists—people whom they view as scientism proponents—reject historical scholarship as a source of knowledge? In the absence of sociological investigation, it is not clear what the alleged proponents of scientism actually believe about this—or even what their accusers believe they believe.

The *OED* definitions may help us to distinguish those fields that count as *sciences* from others that are less theoretical and general in their findings and/or less reliant on postulating falsifiable hypotheses. I doubt, however, that these definitions will enable us to distinguish in any sharp or certain way between the sciences and the humanities. And what about economics, political science, anthropology, and other disciplines within the social sciences? Are these part of science—in a strict sense—or not?

In discussing science's epistemic limitations, McGrath emphasizes its reliance "on the application of observation and experiment in investigating the world" (2011, 77). No doubt the natural sciences do this, but much the same can be said of the social sciences and the humanities. An ancient inscription on a monument is observable, and so is the text of *Macbeth* or *Les Misérables*, or the latest judgment from the High Court of Australia on freedom of political speech. By itself, reliance on observation does little to distinguish science from other fields of inquiry. Moreover, if science is hindered in its ability to answer questions because—*quelle surprise!*—it relies on observation, then much the same applies to the humanities disciplines. I cannot, for example, discover what a newly unearthed royal proclamation states about the events in an ancient battle unless the text is observable by human beings.[4]

To be fair, McGrath also emphasizes experimentation, and here he may be on firmer ground. Nonetheless, it's unlikely that there is a single straightforward criterion for distinguishing science from other forms of serious knowledge production (or for distinguishing scientists from others who make genuine contributions to the sum of human knowledge).

In addition to the definitions that I've cited so far, the *OED* defines *scientific method*: "A method of observation or procedure based on scientific ideas or methods; *spec.* an empirical method that has underlain the development of natural science since the 17th cent." It then elaborates on this at some length, in a way that mentions various approaches to science while tending to emphasize hypothetico-deductive reasoning: the formulation and testing of hypotheses. Although this account has merit, and what it describes is recognizable within scientific practice, hypothetico-deductive reasoning is employed in all areas of human inquiry. Conversely, much work in science is based on close, systematic observation more than on successive conjectures and refutations. Indeed, the *OED* states: "There are great differences in practice in the way the scientific method is employed in different disciplines (e.g. palaeontology relies on induction more than does chemistry, because past events cannot be repeated experimentally)."

The natural sciences tend to rely more heavily on carefully controlled experiments than do the social sciences and—especially—the humanities. However, all these broad fields of inquiry depend, in one way or another, and to some extent or another, on observations and experiments. Thus, Dupré makes a worthwhile point when he doubts that we can distinguish sharply between scientific and nonscientific forms of inquiry, while also emphasizing the importance of distinguishing between projects that genuinely contribute to knowledge (whether or not they are, strictly speaking, "scientific") and spurious projects that misleadingly purport to be scientific (1993, 222). He concludes that science "is best seen as a family resemblance concept" (242).

Likewise, Haack's account of science represents it as continuous with other kinds of serious evidence-based inquiry. It is, however, "more so" in its efforts to overcome human frailties and epistemic disadvantages: hence science's array of observational instruments, its care in contriving and controlling the circumstances in which evidence is obtained (often involving attempts to isolate particular variables), and its conspicuously mathematical character (see Haack 2007, 24–25, 99–109).

There may be no single straightforward methodology that is unique to science, but there are approaches to inquiry that are *distinctively* scientific. This becomes clearer when we consider the early history of Western science. Much pleasure can be found in reading any detailed account of its rise in the sixteenth and seventeenth centuries—such as Stephen Gaukroger's *The Emergence of a Scientific Culture* (2006). Science developed hypothesis by hypothesis, contributor by contributor, and step by step; as it did so, it interacted with much else, such as the broader literary and intellectual culture of Europe.

The great founders and practitioners of the Scientific Revolution were obsessed with thought experiments, actual physical experiments, and each other's ideas. They showed a fascination with the new tools that became available during their lifetimes. These included scientific instruments (such as the telescope) that extended the human senses, precisely crafted experimental equipment, and developments in mathematics. As the sciences took shape, their practitioners were able to engage in systematic study of phenomena that had previously resisted human efforts. These included very distant or vastly out-of-scale phenomena such as those studied by astronomers, very small phenomena such as the detailed composition and functioning of our bodies, and (somewhat later, with the advent of scientific geology) phenomena from deep in time before any human artifacts, buildings, or written records.

Sustained and systematic use of hypothetico-deductive reasoning allowed scientists to test for unseen, hypothesized entities by predicting—and then looking for—their effects on the observable world. Mathematical calculations and models, thoughtfully designed experiments, and innovative experimental equipment aided in the search for those effects. New techniques were developed to isolate experiments from extraneous influences. In all, there is no sharp boundary between science and other forms of rational inquiry, but we can usually agree on which projects are *distinctively* scientific.

A Unity of Knowledge

Accordingly, some fields, and some forms of inquiry, can be more meaningfully classified as scientific than others. The idea of scientism would be clearer, therefore, if it amounted to the claim that distinctively scientific approaches are the best for all fields of genuine inquiry (and that other approaches to obtaining knowledge, such as those normally used by literary scholars and historians, are outmoded and worthless). It is not obvious, however, who makes such a claim.

We seldom, if ever, see such an extreme position taken by reputable scientists, though Massimo's Pigliucci's contribution to this volume includes some examples of scientists who appear hostile specifically to *philosophy*. I must also, to be honest, acknowledge encountering some extreme positions in the informal give-and-take of social-media debates. The evidence here is only anecdotal, but it seems that some of our fellow citizens dismiss the entire practice of humanistic scholarship and interpretation as a bit of a confidence trick. For example, some people appear deaf to tone and nuance in language—and some even seem to deny that these things exist. All the same, let's not attribute crazy positions to individuals without good evidence.

Steven Pinker is one major public intellectual who has, perhaps, left himself unnecessarily open to accusations of scientism. I don't attribute any crazy positions to him, and it's clear that he loves, respects, and understands the humanities, but he makes a very bold claim in an important essay published in the *New Republic* (2013). There, he categorizes René Descartes, Baruch Spinoza, Thomas Hobbes, John Locke, David Hume, Jean-Jacques Rousseau, Gottfried Leibniz, Immanuel Kant, and Adam Smith as scientists. What should we make of this? The efforts of these great thinkers helped lay the foundations of the natural and social sciences, but it requires a long stretch of the imagination to regard them straightforwardly as scientists, or primarily so. Recall that the word *scientist*—when advocated by Whewell and others in the 1830s and 1840s—was not intended to be so wide in its reference. Its meaning in everyday English usage remains essentially unchanged.

Unlike Copernicus, Galileo, Leeuwenhoek, Boyle, and Newton—to name some obvious examples—Descartes, Spinoza, and the others on Pinker's list cannot easily be regarded as scientists *avant la lettre*. Some of them, perhaps most conspicuously Descartes, did contribute to the early phases of the Scientific Revolution, and others (such as Leibniz) assisted in its consolidation. When, however, we think of them as great humanistic thinkers and scholars, that should in no way detract from the rigor of their work, the immensity of their intellectual achievements, or the ongoing impact of their contributions.

In emphasizing this point, I don't necessarily quarrel with the remainder of Pinker's essay. On the contrary, I agree with most of it—but Pinker begins with a provocative assertion about some of the towering figures of modern humanistic learning. His thesis is that science is not the enemy of the humanities, but for many readers his opening gambit may create the opposite impression. Overextension of the word *science*, whose everyday and historical meanings are relatively specific, to cover much that is worthwhile in nonscience disciplines can cause understandable disquiet. Whether or not they are being hypersensitive, some humanities scholars will interpret Pinker with alarm. To them, claiming Descartes, Hume, and the rest for science may suggest that any true intellectual contribution must be part of science as it is popularly understood.

My focus on what distinguishes science from other areas of learning may strike some readers as merely quibbling. If so, I urge them to reconsider. Words communicate, and words used in nonstandard ways can fail to communicate what is intended. A specific claim that *science* can solve such-and-such a problem will tend to convey that the problem is solvable through *distinctively scientific techniques*. It can suggest, furthermore, that contributions from the humanities—or even from such social-science disciplines as anthropology— are unwelcome or irrelevant.

There is much of importance to be discovered, especially with questions about human history and culture, via techniques that cannot meaningfully be classified as part of science. The humanities can of course make use of scientific devices and tests in myriad ways, as when archeologists employ geophysical surveying to locate hidden ruins, or when historians employ carbon dating to confirm the age of documents and artifacts, or even when lawyers search electronic databases for relevant case law. The possibilities are endless. None of this, however, goes anywhere near displacing, as opposed to supplementing and assisting, traditional forms of erudition and scholarship.

Indeed, we should occasionally remind ourselves that our ancestors were not epistemically helpless before the Scientific Revolution. They did have many misconceptions about the world, and such reliable knowledge as they possessed was confined in extent. All the same, scholars from ancient and medieval societies observed events around them, often grasped how other people felt, kept records, translated texts from foreign languages, and developed some impressive knowledge of mathematics. The increasing availability and refinement of distinctively scientific techniques did not prevent scholars from obtaining testimony from others, tracking down lost documents, or making many kinds of ordinary observations. Nothing changed around the seventeenth century (or thereafter) so that traditional scholars could no longer hone the sort of deep knowledge and expertise associated with, for example, mastery of a literary or legal tradition.

Consider a question such as, is the translation of *Les Misérables* on my desk a literal one? There are various ways of investigating this. One approach is simply to ask a relevant expert (perhaps even the translator, if she's alive and approachable). Another is to study French to a level that involves fluency in the language as it was spoken and written in the 1860s. At that point, an able scholar may be able to draw reasonable conclusions by close comparison of the original text with its English translation. Similarly, she might need to master sixteenth-century Italian in order to study Tasso in a scholarly and credible way. At the risk of laboring my point, there is nothing distinctively scientific about any of this.

Permit me, too, some advocacy of the methods used by philosophers. What if we established that some of humanity's most troubling questions—questions with superficially grammatical constructions—are vague, ambiguous, or even (after careful analysis) deeply incoherent? Suppose we make such discoveries about various "transcendent questions" (McGrath 2011, 80, citing views attributed to Sir Peter Medawar) that relate to values, morality, and the meaning of life. Have we been engaging in science? I doubt it: once again, the methods used are unlikely to be scientific in any distinctive way.[5] Nonetheless, something of importance has been discovered, and no supernatural or nonrational ways of knowing were involved.

What should be stated explicitly is that the sciences and the humanities are continuous with each other and can draw on each other's techniques and findings. No cluster of academic disciplines holds a monopoly on particular methods of inquiry. As a result, the boundaries between disciplines and groups of disciplines are blurred. Admittedly, we train different people in different ways in college or university courses, and in technical, professional, and continuing education. The necessity for this relates to variations in human interests and talents—not to mention the limited cognitive powers of any one person and the brevity of human life. However, as and when it is needed to solve a particular problem, individuals with different training can work in teams that utilize their combined skills and knowledge.

Dupré correctly emphasizes that the difference between science and ostensibly "lesser forms of knowledge production" cannot be simply that science is more reliable or epistemically meritorious. He adds, "It might fairly be said, if paradoxically, that with the disunity of science comes a kind of unity of knowledge" (1993, 243). Though they are otherwise very different thinkers, Dupré's conception of a unity of knowledge is not far from Edward O. Wilson's in his more popular book, *Consilience*, published a few years later. Wilson develops the idea of a unity of human knowledge of the world marked by "a deliberate, systematic linkage of cause and effect across the disciplines" (1998, 28).

We should no doubt acknowledge that the rise of modern science was a genuine epistemic advance for humanity. It undeniably *added* something to our available methods of investigation. It did not, however, limit, or subtract something from, our existing repertoire of rational and reliable methods. Significantly, it in no way prevented answers to many questions through methods that were rational and reliable without being distinctively scientific.[6]

Once again, scientific devices and theories can be used to assist humanistic inquiry. Pinker explains, I think convincingly, that this need not entail a vulgar and premature reductionism:

> No sane thinker would try to explain World War I in the language of physics, chemistry, and biology as opposed to the more perspicuous language of the perceptions and goals of leaders in 1914 Europe. At the same time, a curious person can legitimately ask why human minds are apt to have such perceptions and goals, including the tribalism, overconfidence, and sense of honor that fell into a deadly combination at that historical moment. (Pinker 2013)

Pinker avoids the trap of crassly dismissing the methods and language of historical scholarship. As he acknowledges, the natural sciences cannot substitute for "the varieties of close reading, thick description, and deep immersion that erudite scholars can apply to individual works." His acknowledgment of humanistic scholarship is welcome, but we should also accept his point that the humanities can be informed by a more general study of the capacities, limitations, and propensities of human minds.

Scientism and the Theologians

In the opening section of this chapter, I observed that much talk about scientism comes from theologians and religious apologists: it comes from individuals who want to make room for religious belief in a world pervasively influenced by science.

One strategy for them is to argue that science is limited in what it can discover, leaving a role for spookier ways of knowing. I have already referred to McGrath's definition of *scientism* as including the idea that science can tell us about human nature and solve moral problems. The assumption is that knowledge of these things can be obtained from elsewhere, though it's not immediately obvious why that should be so. In any event, some of the information needed to understand human nature may come from the humanities and social sciences. It may, for example, derive from the work of historians and anthropologists.

There is no reason in principle why a variety of disciplines in the natural sciences, the social sciences, and the humanities cannot, between them, give us whatever information is needed. It's also conceivable that a mature intellectual understanding of human nature will elude us indefinitely because of the inherent difficulty of the issues. But if it turns out to be *that* difficult, why imagine that religion (or any supernatural/nonrational way of knowing) can do the job?

McGrath devotes a snarky paragraph to attacking the position within moral philosophy developed by Sam Harris, author of *The Moral Landscape* (McGrath 2011, 79; compare Harris 2010). Elsewhere, I've offered my own detailed (and more temperate) criticisms of Harris (see Blackford 2010) as well as my positive views regarding the central questions of moral philosophy (Blackford 2016). I cannot explore the detail here, but suffice it to say that Harris argues for a form of analytic moral naturalism. That is, he proposes to reduce moral terms such as "(morally) bad" and "(morally) good" to nonmoral language (especially to language that describes the suffering, happiness, or flourishing of human beings and other conscious creatures).[7] Though it faces well-known difficulties, as McGrath correctly points out, analytic moral naturalism remains a respectable view in metaethics. Even if the specific approach taken by Harris is flawed or, as McGrath sneeringly suggests, "lightweight," many philosophers argue for one or another form of analytic moral naturalism. It cannot be dismissed as simply a manifestation of scientism.

Looking beyond this skirmish, we have much to gain, and nothing to lose, from systematic inquiry into the differences and commonalities found in past and present moral systems. We may also be able to establish what, if any, universal psychological elements—desires, inhibitions, emotional tendencies, and so on—underlie the observable phenomenon of morality. Perhaps we can investigate how these evolved and/or how they currently arise from human neurological structures. If so, distinctively scientific inquiry may have much to tell us about morality.

In fairness to McGrath, this sort of knowledge would not establish which, if any, moral system is actually justified. It might, however, help us to understand which kinds of social control are viable (or most viable) for beings like us. The question of justification is notoriously difficult, and it has caused philosophers much in the way of heartache, headaches, and wasted ink. However, one thing seems clear enough: while analytic moral naturalism has its problems, there is also much reason to think that religion cannot give us the answers we want (see Blackford 2016, 79–94).[8]

In *God and the New Atheism*, Haught offers examples of what he regards as science's limits, and he suggests specific questions that he thinks it cannot answer: How do I know someone loves me? How should I understand a work of literature? How do I, or should I, respond to nature? In all these cases, he claims, we should open ourselves, undertake a leap of trust, and make ourselves vulnerable—and so it is, too, with the experience of God (Haught 2008, 44–47). This, however, obscures and mischaracterizes our everyday experience as well as the practice of humanistic scholarship. I concur that the best way, in any particular case, to answer Haught's questions will probably not be (distinctively) scientific. However, these questions are poor analogies for questions about the existence of unobserved, supernatural entities such as the Christian God.[9]

Consider the experience of love. Most of us have substantial innate or learned ability to sense when someone loves us. We can sometimes get it wrong, and this can cause embarrassment, but in the frequent cases where we're correct there will be observable evidence from the other person's demeanor and conduct. We process this rather intuitively, but that makes it no less real. It's complicated, as they say, but it's not supernatural.

We may experience the world around us as beautiful, sublime, or astonishing, although that is not primarily a cognitive experience. But at any rate, nothing supernatural is involved when we respond to the beauty of Uluru or the Grand Canyon or an ancient, magnificent forest—or perhaps to the delicacy of a flower or a spider's web. Our senses reveal natural phenomena to us in an ordinary, familiar way, and they provide us with relevant knowledge of what we're experiencing.

Interpreting works of literature is a different matter again, and the methods used by literary critics (or by ordinary readers with well-honed sensitivities) are remote from those used by astrophysicists, microbiologists, or experimental psychologists. Skilled literary interpretation involves, among other things, a mastery of the nuances and registers of the relevant language in which the text was composed, familiarity with literary genres, modes, and tropes, and a deep understanding of the relevant traditions of composition and reception. Literary interpretation is not, however, a practice that transcends observable evidence.

Perhaps there is no single objectively authoritative interpretation of a complex literary text—let's again say *Les Misérables*. More broadly, the humanities are hampered in answering many of the questions that they set themselves: questions about the meanings of artistic works, for example, or questions about the perceptions and goals of political leaders. Often, such questions cannot be answered decisively on the available evidence—but acknowledging this is not the same as claiming that anything goes in humanistic scholarship.

There are explicable reasons to maintain that Jean Valjean is *not* the villain of *Les Misérables*, and that Mary Shelley's *Frankenstein; or, The Modern Prometheus* is *not* a polemic in favor of creating artificial megafauna in the lab. (If you disagree here, either something has gone badly wrong with your understanding or you're an unashamed contrarian.) There are important facts about, for example, the languages in which these texts were written, the words employed in particular sequences, and the historical and artistic contexts of composition. These facts make some interpretations richer, more coherent, and (quite simply) more plausible than others. They also rule out many potential interpretations.

Distinctively scientific means do not, then, provide the best approach to every question. At the same time, science is not limited in a sense that lends credence to supernatural or nonrational ways of knowing. If admirers of these want to justify them as reliable, they will need to do better than complaining about their opponents' propensities for scientism. For example, they could try to demonstrate that their preferred techniques have a convincing track record of producing checkable knowledge. We can await that demonstration with interest. While we're waiting, neither the sciences nor the humanities have access to reliable supernatural methods, but insistence on this point should not be denigrated as scientism.

When Science Met Religion

Talk about the evils of scientism is often employed by religious apologists who would have us believe that the success of science leaves the credibility of religion unscathed. They resist science's tendency to disenchant the world. Christian apologists, in particular, often go further by suggesting that Christianity even assisted the rise of modern science (see, for example, Plantinga 2011, 265–303).[10]

This is a fascinating question, but I cannot deal with it here in any detail. In *The Emergence of a Scientific Culture*, Gaukroger examines European intellectual history from the rise of neo-Aristotelian natural philosophy in the thirteenth century to developments in science in the late seventeenth century. He expressly does *not* argue that Christianity, or medieval Christian theology, provided the impetus for sixteenth- and seventeenth-century science. However, he develops an impressive case for a much narrower thesis: that in the 1680s and 1690s, especially in England, there was a widespread view that natural philosophy could be used as a source of evidence for God (Gaukroger 2006, 505). As a result, certain strains of Christian theology deserve some credit for the *consolidation* of science that took place in Europe in the late seventeenth century and thereafter.[11]

But even if some tendencies in medieval and/or early modern Christian theology provided science with fertile ground, science, in turn, possessed a latent tendency to inspire doubts about Christianity itself. As it developed in power and confidence, science challenged many precious beliefs about the world, including beliefs that relate to human exceptionalism. Its ongoing achievements have suggested the possibility that all the phenomena of nature can (at least in principle) be explained without appeals to supernatural entities. The problem that emerged within Western modernity for religious (and specifically Christian) worldviews was not so much that science disagreed with them on specific points, although it often did so. More significantly, the old explanatory systems were revealed as premature and—by science's standards—poorly evidenced.

During and since the Scientific Revolution, advances took place in numerous fields, producing the sense of a comprehensive transformation of human knowledge. Science developed a radically new, and increasingly compelling, description of the cosmos, our place within it, and ourselves. In that sense, it has become our best, and indeed only, guide to the overall reality of space and time. Previously, human beings were greatly confined in how much of the world they could even begin to understand. Notwithstanding the abilities of scholars in prescientific societies, they possessed little knowledge of the cosmos except on human scales.

Science may not outright disprove the existence of a God or gods, but its progress has done much to make atheism and irreligion more thinkable. There have, however, been other influences on the erosion of religion, including influences from the humanities. During the eighteenth and nineteenth centuries, textual-historical study of the biblical texts cast doubt on traditional understandings of their origins, mode of composition, and historical reliability. At the time, this was a devastating blow to the credibility of the churches.

If Christianity—or religion more generally—could establish some claim to epistemic authority stemming from its access to nonrational ways of knowing, this might greatly shore up its damaged credibility. Unfortunately for theologians such as McGrath and Haught, their arguments go nowhere near establishing such a thing.

Conclusion

Over the past four to five centuries, there has been an explosion of knowledge in the natural sciences, the social sciences, and the humanities. It continues today, with no end in sight. We have learned enough to lose trust in many traditional belief systems, but much still remains to be discovered and we do not have a finished picture of the natural world or of human history and culture. By now, we can safely assume that no supernatural entities or forces will be discovered by the sciences or the humanities in their continuing inquiries. The track record of the natural sciences in particular lends support to philosophical naturalism as a minimalist worldview, but it is not a completed and comprehensive one.

Some scientific findings are most unlikely to be revised, since the evidence is overwhelming. In other fields of inquiry, the evidence is incomplete or more ambiguous. In such cases, there may be considerable likelihood of revision, and sometimes we'd do best to suspend judgment about what to believe. In fields of inquiry that push few emotional or political buttons, the incompleteness or ambiguity of available evidence may not be very troubling: we can acknowledge the situation, lay out some interpretations that are consistent with the evidence we have, and consider what more we can do to narrow the options. In hot-button fields—often but not always in the humanities—that is not usually how the situation plays out. In the absence of decisive evidence, these fields can become intractably divided along political and moral lines.

I am not advocating any radical skepticism about the humanities, which indeed can produce knowledge, but we often have reason for caution about their claims. Good research in the humanities requires a practical mastery of languages and other human-made systems of concepts and meanings (such as traditions of art or philosophy or law). Genuine mastery deserves respect. Indeed, practitioners within the humanities can do themselves a disservice by insufficiently emphasizing the hard-won erudition and expertise required to solve problems of interpretation and to make intellectual progress. As it appears to me, however, much of what currently passes for literary and cultural criticism falls strikingly short of this mastery.

Thus, scholars in the humanities have some legitimate concerns about unfavorable intellectual trends and attitudes. But if they want their disciplines to remain intellectually credible and socially useful, they had better keep their own houses in order.

Throughout this chapter, I have argued that we can investigate many questions—often important ones—without doing anything distinctively scientific. But this in no way supports the idea that there are reliable ways of knowing that are radically discontinuous from ordinary observation and reasoning. Nonrational methods such as faith and revelation do not penetrate further than the forms of inquiry used by natural and social scientists, by humanities scholars, and by ordinary people solving problems in their daily lives.

I have acknowledged an intellectual trend toward glorification of technoscience. However, the term *scientism* is not a useful one for this. For a start, it falsely assumes that glorifying technoscience is always a bad thing. Why should it be?[12] More crucially, the word carries too much theological baggage—as seen in its *OED* definition, its history, and its current weaponizing by theologians and religious apologists. I propose, therefore, that the word *scientism* be shelved.

Nonetheless, there remain genuine issues about contemporary intellectual trends and the epistemic reliability of various disciplines and methods. Alas, I doubt that there is a good umbrella term for the issues I've identified. They include mathematics fetishism; inappropriate mimicry (by the humanities) of the superficial trappings of science; disquieting extensions of the word *science* beyond its popular and historically settled meaning; varieties of science denial and unwarranted hostility to science; and various kinds of philistinism, cynicism (whether about the sciences or the humanities), intellectual vulgarity, and politicization of research. Though *scientism* is not one of my words, we do need a vocabulary, preferably a rich one, to scrutinize unfortunate intellectual trends and their possible connections. I hope that this chapter—not least this final paragraph—provides assistance.

······

Notes to Chapter One

1. While there are diverse accounts of faith in Christian theology, prominent among them is a conception of faith as a nonrational way of knowing. For more detail, see Blackford and Schüklenk 2013, 131–42.

2. Note that the *OED* also has a definition of *scientific method*—I'll return to this.

3. In the usage of the time, a naturalist was basically a student of plant and/or animal species.

4. Some research carried out in the humanities—especially in philosophy—is largely conceptual. That is, it analyzes ideas and logical relations. But scientists can engage in this, too. The fact remains that any epistemic limits applying to science also apply to the humanities.

5. I don't intend here to rule out the methods of so-called experimental philosophers. The point is merely that the more traditional methods used in analytic philosophy are rational (and have some prospect of being epistemically and socially valuable) without being distinctively scientific.

6. The progress of science may of course have done much to undermine various *unreliable* and *not-so-rational* methods that were in use.

7. My criticisms of *The Moral Landscape* include what I take to be its implausible efforts at reductive moral semantics (see Blackford 2010, 57–60; compare Kalef, this volume).

8. I doubt that analytic moral naturalism can provide the whole story about morality, and (to simplify drastically) I favor an error theory about at least some of our everyday moral language. However, none of my arguments (in Blackford 2016) draw on supernatural or nonrational ways of knowing. They rely, in the main, on observation and analysis of relevant social institutions, concepts, expressions, and words.

9. Elsewhere, Udo Schüklenk and I discuss Haught's examples at greater length, and with a slightly different emphasis (Blackford and Schüklenk 2013, 154–56).

10. Such claims do not always come from religious apologists. Even Edward O. Wilson, a somewhat half-hearted deist, has expressed sympathy for the idea that Christianity nurtured modern science (1998, 32).

11. He elaborates further in Gaukroger 2010.

12. By all means, let's celebrate medical cures, wonderful space vehicles, new discoveries about the dinosaurs, and much more—make your own list!

Literature Cited

Blackford, Russell. 2010. "Book Review: Sam Harris's *The Moral Landscape*." *Journal of Evolution and Technology* 21, no. 2 (December): 53–62, http://jetpress.org/v21/blackford3.pdf.

———. 2016. *The Mystery of Moral Authority*. Basingstoke: Palgrave Macmillan.

Blackford, Russell, and Udo Schüklenk. 2013. *50 Great Myths about Atheism*. Malden, MA: Wiley-Blackwell.

Dupré, John. 1993. *The Disorder of Things: Metaphysical Foundations of the Disunity of Science*. Cambridge, MA: Harvard University Press.

Gaukroger, Stephen. 2006. *The Emergence of a Scientific Culture: Science and the Shaping of Modernity, 1210–1685*. New York: Oxford University Press.

———. 2010. *The Collapse of Mechanism and the Rise of Sensibility: Science and the Shaping of Modernity, 1680–1760*. New York: Oxford University Press.

Haack, Susan. 2007. *Defending Science—within Reason: Between Scientism and Cynicism*. Amherst, NY: Prometheus Books.

Harris, Sam. 2010. *The Moral Landscape: How Science Can Determine Human Values*. New York: Free Press.

Haught, John F. 2008. *God and the New Atheism: A Critical Response to Dawkins, Harris, and Hitchens*. Louisville: Westminster John Knox Press.

McGrath, Alister. 2011. *Why God Won't Go Away: Engaging with the New Atheism*. London: Society for Promoting Christian Knowledge.

Oxford English Dictionary. 2014. 3rd ed. (online version).

Pinker, Steven. 2013. "Science Is Not Your Enemy." *New Republic*, August 6, http://www.newrepublic.com/article/114127/science-not-enemy-humanities.

Plantinga, Alvin. 2011. *Where the Conflict Really Lies: Science, Religion, and Naturalism*. Oxford: Oxford University Press.

Wilson, Edward O. 1998. *Consilience: The Unity of Knowledge*. London: Little, Brown.

2

Plus Ultra:
Why Science Does Not Have Limits

MAARTEN BOUDRY

Introduction

In Greek antiquity, the promontories flanking the Strait of Gibraltar were known as the Pillars of Hercules. Erected by the eponymous hero as part of his tenth labor, they marked the end of the known world. According to Renaissance tradition, the pillars bore the inscription *nec plus ultra* (no further beyond), serving as a warning to intrepid sailors. No one could tell what, if anything, lay beyond. Francis Bacon used this allegory as the frontispiece to his ambitious program for a new science, the *Novum Organum* (1620); it depicts a ship sailing through the mythical pillars, venturing into the Unknown.[1] Ever since Bacon's campaign for a "new philosophy," scientific knowledge has increased at a breathtaking pace. In a matter of four centuries, we have unveiled the origins of our species and all other life on the planet, explained the nature of infectious diseases, landed on the moon, split the atom, and (just recently) detected gravitational waves.

Not only has science accumulated ever-more knowledge but it has also drastically increased its scope, from the outer reaches of the universe to the inner sanctum of the human mind. Many people don't welcome this meddlesome scientific curiosity. Whether you like it or not, science has reached pretty definitive answers on many questions of human interest, leaving traditional myths and doctrines in ruins. In the wake of this impressive body of knowledge, and the undeniable technological fruits that it bears, an educated person living in the twenty-first century can no longer reject the authority of science altogether. But is the sky the limit? Many people maintain (or hope) that there must be some limitation to scientific investigation. Science is a powerful instrument, they admit, but it cannot answer every question. Surely there must be some mysteries that science will never unravel, some domains where it cannot venture? Science can launch rockets and split atoms, but it cannot tell us why we are here. Or perhaps questions of origins are within its purview after all, but the human psyche remains off-limits. Or if not the whole of our psyche, then at least consciousness, love, or morality. And in any case, is not much of human knowledge arrived at through nonscientific means?

According to these people, for science to venture beyond its natural limits is to commit the sin of scientism. Science itself is not the problem, they maintain, but rather the overextension of science into all domains of inquiry. Traversing the Pillars of Hercules is a form of scientific imperialism, hubris, fundamentalism, or totalitarianism.

But what are those limits of science? Are there other, equally valid ways of knowing besides science, or do the limits of science coincide with the limits of human knowledge? Should we imagine science as one sort of vessel, equipped to navigate certain waters, but perhaps not others? If science has identifiable limits, are there other ways of knowing that can venture beyond?

The Varieties of Scientism

There is no consensus on the meaning of the term *scientism*. In this chapter, I will focus on its (primary) epistemological sense, relating to the limits of scientific knowledge, but there are a number of other usages. It may refer to an excessive deference to science (Haack 2012; Sorell 2002), an unhealthy obsession with Grand Unifying Theories, or a bad habit of denigrating disciplines other than the natural sciences (for examples, see Pigliucci, this volume). Other people use the term to criticize overblown confidence in the future progress of science, for example the idea that science will soon make us all immortal (see for example the futurist Aubrey de Grey). The most crucial aspect of scientism, according to Sorrell's book on the subject, is "the thought that the scientific is much more valuable than the non-scientific, or the thought that the non-scientific is of negligible value" (2002, 9). But I will not discuss these forms of overheated or misguided science enthusiasm. It is quite clear that even our best science remains fallible, and that there are other valuable things in life besides science: art, literature, sex, entertainment, gardening, perhaps even sports.

It should come as no surprise that the term *scientism*, in its epistemological sense, has many abusers. People of all stripes are quick to trot out the scientism gambit whenever some scientific theory encroaches on their turf or threatens their particular worldview. Mediums and psychics use it to ward off inquisitive minds (Law 2011). Religious believers hurl it around to protect sacred doctrine from the advances of science (Boudry 2015; Clark 2015). Postmodern relativists press it into service to "unmask" the pretensions and imperialist ambitions of science, bringing it down to the level of other, equally valid ways of knowing. Even some philosophers and humanities scholars seem overly anxious that zealous scientists are bent on a hostile takeover of their discipline, and have tended to use the word *scientism* in a defensive and poorly justified fashion (Wieseltier 2013; Pinker 2013).

In short, whatever value there is to the notion of scientism, the term's real meaning all too often boils down to "science I don't like." For that reason, the very mention of the term in scientific circles often provokes eye rolls and groans of impatience. In an ironic twist, some ardent supporters of science have started to embrace the term as a defiant nom de plume, as a way of preempting their critics (Ladyman and Ross 2007; Rosenberg 2011). And yet, the question of scientism deserves an honest appraisal. We should not leave it to theologians and postmodernists to pontificate about the limits of science. Yes, the term has plenty of abusers, but any concept with a normative value carries the potential for abuse. Take *pseudoscience*, an inherently pejorative term that, to the best of my knowledge, has not been taken up as a badge of honor by anyone yet. In one of his intelligent design tracts, the Berkeley professor of law Phillip Johnson opined that "Darwinism is a pseudoscience that will collapse once it becomes possible for critics to get a fair hearing" (2002, 141). Does this hijacking of the term *pseudoscience* mean that we should abandon the concept altogether (Pigliucci and Boudry 2013)? Of course not.

In this chapter, I will retain the pejorative sense of *scientism*, as opposed to the positive sense as a badge of honor.[2] By definition, then, scientism is that which pushes science beyond its epistemological limits. If there are no such limits, then there is no such thing as scientism. First, I will defend the naturalist (holistic) conception of the web of human knowledge, according to which science is interwoven with everyday knowledge, philosophy, and other academic disciplines. Drawing inspiration from evolutionary epistemology, I argue that what we now identify as the borders between folk knowledge and science is contingent on our cognitive makeup, and thus is largely an accident of history and biology. From this holistic point of view, the notion of limits to science becomes difficult to defend. Still, I will try to find out if there are any clean "breaks" in our web of knowledge, which even an inveterate naturalist should be willing to accept. Working by elimination, I discuss two realms that are often regarded as necessarily beyond the purview of science: the supernatural realm and the moral realm. In the case of the supernatural, I defend the inclusive nature of science, attacking the methodological strictures imposed on science by those who want to reconcile it with religion. In the case of morality, finally, I argue that science reaches some sort of limit. But this, as we will see, still provides no succor to advocates of "other ways of knowing."

Everyday Knowledge and the Manifest Image

Let's start with a completely unsuspected source, who is not an enemy of science by any stretch. According to the philosopher Hilary Putnam, it is of utmost importance to acknowledge that there are other forms of knowledge besides science: "A view of knowledge that acknowledges that the sphere of knowledge is wider than the sphere of 'science' seems to me to be a cultural necessity if we are to arrive at a sane and human view of ourselves or of science" (2013, 5). But how are we to make sense of such limits? It all depends on our definition of *science*. If we have a rather narrow definition, there will clearly turn out to be other sources of knowledge besides science. But if we adopt a more inclusive definition of science, the question of the limits of science will receive a completely different answer. As the entry for *scientism* in the *Dictionary of the History of Science* puts it, "The legitimate scope of science is of course at issue: one man's science is another man's scientism" (Bynum, Browne, and Porter 2014, 381). For example, if we restrict the concept of science to the natural sciences, excluding history and the humanities, no sane person would dispute that there are forms of knowledge besides science (see Rosenberg, this volume, for a counterpoint). If, however, we define science so broadly as to encompass any conceivable form of human knowledge, we will have defined any limits to science out of existence. We have to steer between Scylla and Charybdis, on pain of trivialization.

From a naturalist (holist) point of view, human knowledge forms a tightly interwoven web (Quine 1951; Haack 2007).[3] What we call science is enmeshed in that web, but there is no sharp distinction between science and other forms of knowledge, and the strands within the web are mutually interdependent. As a case in point, consider the borders between science and everyday knowledge. Putnam notes that there is plenty of everyday "practical knowledge" that can't be "scientized," for instance when we pick up the meaning of a word in a foreign language by hearing how it is used in context. Such inferences, according to Putnam, "are not, in any serious non-trivial sense, scientific inferences" (2013, 70).

There is of course some value to Putnam's distinction. Some facts are plain for everyone to see and do not demand any special expertise or theoretical knowledge (e.g. facts about tables and chairs, animals and plants, rivers and rocks). Others demand technical expertise, but are of a more local and applied nature (e.g. plumbing, gardening, violin making). Still others demand theoretical knowledge, but of the kind that every normal human being is naturally endowed with (e.g. language skills, folk psychology, folk physics).

Science, as commonly understood, is concerned with forms of knowledge that are not obvious, intuitive, or immediately accessible. The problems that keep scientists busy demand technical apparatus, statistical analysis, careful measuring and cross-checking, controlled observation, and extensive collaboration. But how important is that distinction? Our particular perceptual and cognitive endowment determines what knowledge belongs to the "manifest image" (Sellars 1963), and what is left to specialized scientific investigation or craftsmanship. For example, humans need science to detect UV light, but bees do not. If we could see Jupiter's moons with our naked eyes, we would not have needed a scientific genius like Galileo (and some fancy equipment) to discover them, and we would not have called this knowledge scientific. If our eyes had built-in microscopes that allowed us to see bacteria and detect infections, we probably would not be talking about the germ *theory* of disease. Bacteria would just be part of our "manifest image," our familiar furniture of the world.

If creatures from the Proxima Centauri system were to visit our planet, much of human behavior might initially appear totally incomprehensible to them. Suppose they observe one human being punching another one in the face, after the mouth of the first one had opened and his vocal cords started vibrating. For any human witness, it would be perfectly obvious that the second human being became angry after the first insulted his mother. For a Centaurian scientist, however, not endowed with an intuitive understanding of the human emotional repertoire and our intentional psychology, such behavior might appear completely baffling. As Dennett (1981) has argued, the real patterns in human behavior would not become visible for the Centaurians until they learned to treat humans as intentional systems with beliefs, desires, and emotions. Once they have adopted this "intentional stance," the Centaurians might then proceed to conduct a thorough scientific study of human facial expressions and body language, human emotions related to family honor, game theories about emotional commitment, and the psychology of deterrence and retaliation.

Much of what we human beings get for free in the form of folk psychological knowledge would be state-of-the-art science for Centaurians, obtained by carefully observing regularities in human behavior, testing hypotheses about our dispositions and reactions, and so on. Our alien friends, in turn, might have innate cognitive and perceptual talents of their own, yielding knowledge that, in human beings, would require the use of cutting-edge science. Perhaps they could directly observe red-shifts in the stars, or pick up curvatures in space-time the way our vestibular organs pick up the direction of gravity.[4] Questions about the "limits" of science lose force here.

Let's now return to Putnam's example of understanding a foreign language. Does his argument show that such knowledge is off-limits to science? At most, it shows that we humans do not *need* science to obtain certain forms of practical knowledge, because nature already equips us with serviceable language skills by the time we reach adolescence, without much explicit instruction. Language skills come "for free," with most of the research and development already carried out by blind evolution. In most everyday contexts, our language modules perform admirably well and allow us to figure out the meanings of foreign words. But that does not entail that the study of foreign languages is off-limits to science. Indeed, professional linguists conduct *scientific* investigations into the syntax, semantics, and pragmatics of different languages. Psycholinguists reconstruct the unconscious inferences and heuristics that underlie our innate language skills, and evolutionary psychologists try to figure out how these modules have evolved and are developmentally implemented. Though it is true that we normally don't use science to learn a new language, we do use scientific methods to solve more daunting problems of translation, for instance to decipher an ancient text in an unknown language.

In short, what Putnam gestures at are not the epistemological limits of science but the conditions under which we need it. This, however, is contingent on both the cognitive makeup of our species and the standards of evidence and explanatory depth we demand in any given situation. I don't need a scientist to tell me that I had some yogurt for breakfast today, but I do need a nutritionist to tell me its caloric value, or to tell me what Belgians in general eat for breakfast. I don't need a scientist to tell me if it's raining right now, though I do need a meteorologist to tell me if it's going to rain tomorrow. I don't need a scientist to predict what will happen if I insult someone's mother, but I do need a scientist to tell me if punching this person in the face really will help me to vent my anger (it won't—it will make me more aggressive; see Tavris 1988).

Nonscientific knowledge, by the way, can be descriptive or predictive, specific or more general, applied or theoretical, just like scientific knowledge. My plumber may be quite adroit in investigating leakage, but I would not ordinarily call him a scientist.[5] Nor would we call an expert violin maker a scientist. From an epistemic point of view, however, there are plenty of commonalities between what a biologist is doing in the lab and what a plumber is doing when he is trying to locate a leak in my water supply. The plumber is making observations, testing out different hypotheses, using logical inferences, and so on. The main difference is that he is working on a relatively mundane and isolated problem (my sink), which is both simple enough to solve on his own with reasonable confidence, and parochial enough to concern no one but me and him—no need to attend conferences and submit articles to peer-reviewed journals about my kitchen. It would certainly be a peculiar usage of language to call my humble plumber a scientist, but then again, it would be strange to think that any point of epistemological interest hinges on withholding that status from him.

Are Inuit hunters tracking narwhals scientists? Not really, but their sleuthing techniques are so impressive that real scientists have enlisted their help to study these elusive animals.[6] As Thomas Huxley observed, "Every time a savage tracks his game he employs a minuteness of observation, and an accuracy of inductive and deductive reasoning which, applied to other matters, would assure some reputation as a man of science" (1871, 18). And as Susan Haack puts it, "Scientific inquiry is continuous with everyday empirical inquiry—only more so" (2007, 94). The most striking differences between hunter-gatherers and scientists—technical apparatus, systematic observation, peer review, statistical analysis—are not the most important from an epistemic point of view. In a similar fashion, what paleontologists were doing when they were figuring out the cause of the extinction of the dinosaurs at the end of the Cretaceous bears a striking resemblance to what a detective is doing when looking for a smoking gun (Cleland 2002). The problem of the extinction of the dinosaurs is far too daunting for a single human researcher to crack, given our cognitive and perceptual talents, but that does not make it of a different order altogether. It definitely does not show that tracking animals or solving murder mysteries is somehow beyond the limits of science.

Quite apart from the continuity of science with everyday knowledge acquisition, it should be obvious that science would not even get off the ground without everyday knowledge (but see Peels, this volume). As Putnam writes, the very practice of science needs to presuppose some forms of nonscientific knowledge. As far as we can tell, all scientific work in the universe is performed by human beings, and is thus constrained by their cognitive and perceptual talents. Even the remotest and most abstruse objects of scientific investigation are accessible only through our innate perceptual affordances: seeing specks under the microscope, reading off a dial, digging up shards or fossils. Knowledge about those primary sense-data is not by itself scientific— any fool can read a dial, peer through a telescope, or see an excavated fossil.[7] But again, if that constituted a limit to science, our question about scientific limits would become completely trivial. Science is not a divine gift descended upon us from the heavens. It was gradually built up out of ordinary human cognition—but we knew that all along (Thagard 2012; Sterelny 2010).

Plumbers, piano makers, and paleontologists all possess specialized factual knowledge, but the differences between their methods are not indicative of any epistemological limit to science. Science, in the narrow sense, is just a convenient word to single out the collective human endeavor for developing a more systematic understanding of reality, and for ferreting out those parts of the universe that do not lend themselves to immediate observation or folk understanding. Lots of human knowledge falls outside science in this sense, but not in a way that establishes any epistemic limits to science. If there are any limits between science, craftsmanship, and everyday knowledge, they are of a pragmatic nature, reflecting our cognitive makeup, our human interests, and our epistemic division of labor.

Academic Fences

If the borders between science and everyday knowledge are porous and pragmatic, the same is true for the borders with neighboring disciplines in academia (for a fuller treatment, see Blackford, this volume). A university campus map need not track any epistemic fault lines. I have previously argued that we should distinguish between two demarcation problems in philosophy of science: the *normative* demarcation between good science and pseudoscience, and the *territorial* demarcation between science and other academic endeavors, such as philosophy, history, or mathematics (Boudry 2013). Normative demarcation is a pressing matter, as much hinges on where a theory falls. Pseudoscience is harmful to society, and philosophers of science had better be able to separate it from the real thing.[8] Unlike the normative science–pseudoscience divide, however, territorial demarcation lines carry little epistemic import. Regardless of whether you include history or philosophy within the domain of science or not, there are no dire consequences for these disciplines. One indication that territorial demarcation is a largely semantic issue is that the corresponding word for "science" does not have the same scope in different languages. The German term *Wissenschaft*, for instance, is considerably broader than the English *science*, comprising the humanities as well as the natural and social sciences.

Take the example of history. Historians use methods that are appropriate to their subject domain, such as thick descriptions and multilayered causal narratives. Does that mean that there is a unique "historical method," different from the methods used in the natural sciences? Perhaps, but not in the sense that any interesting limit to science has thereby been demonstrated. Many people who think that there is a sharp distinction between history and the natural sciences, because historians "tell stories" and do not produce predictive knowledge, often have a simplistic and outdated conception of the natural sciences in the first place (see Ross, this volume).[9] Moreover, history and the natural sciences are mutually dependent on each other's findings and methods (Pinker 2013). For instance, historians rely on a body of physical knowledge about radioactive decay rates to date historical documents and artifacts. In their turn, historians can provide data to linguists and cognitive scientists who study the evolution of human language.

Or take my own discipline: philosophy. Some of my colleagues still regard philosophy as independent from or conceptually prior to science. Philosophers supposedly have their own proprietary problems and methods, and if they need to be bothered with scientific knowledge at all, it is merely to critically *question* its assumptions and presuppositions. In the eyes of these philosophers, it is a serious category mistake to conflate a purely philosophical problem with a "merely" empirical one. These ideas, while still fashionable in some philosophical quarters, are on the way out (I may indulge in some wishful thinking here). Much of philosophy is now tightly ensnared in the web of knowledge. Philosophy of mind shades into cognitive science, neurology, and linguistics. Epistemology is intertwined with cognitive psychology and evo-lutionary biology. The special sciences often deal with conceptual issues that can be characterized as broadly "philosophical" in nature, and to which phi-losophers have indeed made useful contributions. In biology, for instance, the study of systematics shades into philosophical debates about natural kinds, essentialism, and family resemblance (Hull 1990). Philosophers of mind have made important contributions to the scientific study of the human mind: for instance Dennett (1993), with his *multiple drafts* model of consciousness and his work on the *intentional stance* mentioned above, or Fodor (1983), with his seminal work on the mind's modularity. Contemporary psycholinguistics is indebted to philosophy of language: for instance the analysis of speech acts and conversational implicatures (Grice 1989; Austin 1962). Human knowledge is a tight-knit web, and there is no way to disentangle science (no matter how you construe it) from such neighboring disciplines as history, philosophy, or even metaphysics (Ladyman and Ross 2007). If it's not entangled in the web of knowledge, chances are that it's not knowledge at all.

From this unabashedly naturalist and holistic vantage point, if science has any limits deserving of that name, it must be with some way of knowing that is sufficiently discontinuous with science to make for a (relatively) crisp border. Scientism, then, would be an attempt to overstep that boundary, to apply the methods of science in a domain where they have no authority. In the following section, I consider two candidate domains that are often claimed to be inaccessible to science, and that seem to make for a sharp boundary: the supernatural realm, and the moral realm.

The Supernatural Realm

Do the methods of science permit us to obtain knowledge about supernatural and paranormal entities? It would be easy to dismiss this question out of hand as follows: no supernatural realm exists, and that which does not exist cannot possibly be off-limits to science. (As in the classic conundrum: is the present king of France bald?) Even for a staunch atheist, however, it might be inter-esting to reflect on the question: if some god did exist, would (s)he be on the radar of science? A number of philosophers and scientists, unbelievers as well as believers, have argued against this notion (Pennock 1999; Forrest 2000). Science, they maintain, cannot deal with supernatural claims as a matter of *principle*. This doctrine goes by the name of methodological naturalism.

In earlier publications, my coauthors and I have criticized this methodological stricture, arguing that science does have a bearing on supernatural claims (Boudry, Blancke, and Braeckman 2010, 2012). If there really was some sort of supernatural being at work in this universe, intervening from outside or influencing events in some other mysterious way, what would prevent scientists from smoking this being out? As long as this god or ghost causally interacts with the observable world, (s)he/it would be amenable to scientific investigation. Consider extant religious traditions. Depending on the specific creed, supernatural beings have different degrees of involvement with the natural world (e.g. creations, revelations, miracles, answering prayers, etc.), but although many gods are somewhat shy and aloof, usually they are not *completely* separate from the natural realm.[10] Many supernatural claims enjoying wide appeal today, among both laypeople and academic theologians, are in fact perfectly amenable to scientific investigation (Fishman and Boudry 2013; Edis 2002). Even a deist God could potentially leave traces, producing a universe different from what one would expect without a divine igniter.

While it may be hard to imagine, at this point, how supernatural beings could make their comeback in science after being gradually and completely expunged, there is nothing that prevents this in principle. Suppose dowsers struck gold time and again, or petitionary prayer were found to work, or we discovered that invisible spirit beings are lurking in the woods. Would we have reached the limit of science? Would scientists throw their hands up in the air and admit defeat? Exactly the opposite would happen. There would be a flurry of excitement in the scientific community, and researchers would immediately try their best to figure out what was going on. In the case of dowsing, geologists would start to investigate the conditions under which dowsing works. Must the dowser possess special skills, or does the magical power inhere in the dowsing rod itself? What is the maximal radius of dowsing rods? If we cut the rod in two, do we now have two functional dowsing rods? What kinds of materials are the rods sensitive to? Does this mysterious dowsing skill fit within the fold of physicalism, or will it uproot some of our deepest metaphysical assumptions about the world?

Granted, at some point such scenarios of supernatural discoveries start to look slightly preposterous. It is not even clear whether supernatural concepts are coherent in the first place. (If ghosts are immaterial and can move through walls, how come they can rattle shutters and doors?)[11] But if some supernatural views are too vague and incoherent to even be empirically evaluated, that is not a limitation of science but a failure of those who dabble in such views. In any case, if we were to find strong evidence for the supernatural, scientists would have to abandon their provisional commitment to methodological naturalism. Those who see the methodological naturalism of science as an *intrinsic* limitation fail to appreciate its open-ended and inclusive character. In science, findings feed back into methods, and nothing is carved in stone forever. For example, the double-blind method in medical research became standardized because we discovered the unconscious influences of expectation and the power of the placebo effect. If our psychological constitution were immune to such unconscious influences, our research methods would follow suit. If Centaurian psychology does not exhibit anything like the placebo effect, research protocols for Centaurian medicine will not and need not include the double-blind feature. In that case, ailing Centaurians can just swallow pills (or whatever suits their anatomy) and wait to see what happens.

Methodological naturalism is a pragmatic guideline, not a precondition of science (Carroll 2016). In different disciplines and at different times in the history of science, scientists have learned that they do not make any headway by invoking miraculous causes and supernatural beings. Supernatural explanations for the phenomena of nature, for example in terms of ghosts or gods or goblins, have been superseded by blind and impersonal causes (Boudry, Blancke, and Braeckman 2010; Edis 2002). Because many people mistake this naturalistic outlook of science as a self-imposed limitation, they think that the methods of science *prohibit* it from evaluating supernatural claims. To venture beyond the natural realm, so the argument goes, you need other ways of knowing. But the absence of gods and ghosts from science is just a contingent result of our investigations into the world, and the dismal track record of supernatural explanations.

Similar points apply to other methodological principles of science. For instance, if you think that the practice of science *presupposes* the lawful regularity of nature, or the comprehensibility of the world, or the usefulness of mathematics, you are getting things exactly backward (Fishman and Boudry 2013). Scientists have *discovered* regularities in the world, and have learned that mathematics (some of it, not all) furnishes remarkably useful tools for understanding reality. But mathematics is not an indispensable tool of science (Darwin's *On the Origin of Species* does not contain a single equation). Historians and sociologists have not discovered anything like the "laws" of human societies, nor do they need to, despite the misconceptions of Marx, Hegel, and others (Popper 1957).[12]

The open-ended nature of science makes it hard to construe claims about its limits. There is no circumscribed set of scientific methods independent from the contingent results of science and the features of our cognitive makeup. New scientific findings can inspire new methods, and these are subsequently incorporated in the toolbox of science. It is a form of Whig history to take for granted that the way science is conducted now was somehow inevitable, or how science *ought* to be.

Does this mean that science can venture anywhere? Not necessarily. For example, it may be argued that *some* conceivable supernatural beings would be beyond the limits of any science, human or otherwise. Suppose there is a god, but one who abides by a strict noninterventionist policy, never offering himself up for observation, and being infinitely devious in covering up his tracks. In such a world, indeed, science would be impotent to find out anything about this god. But we should be perfectly happy to acknowledge that hermetically untestable claims, such as Bertrand Russell's notion that the world was created five minutes ago, with perfect appearance of old age (Russell [1921] 2005), are in some sense beyond the limits of science to evaluate (which is of course no reason at all to take them seriously).[13] A partial analogue can be found in the real world: events that fall outside the "past light cone" of earthly observers are inaccessible to science, because any information coming from those points in space-time would have to travel faster than light to reach us, which is prohibited by the theory of special relativity.

But this is no reason for the faithful to rejoice. If science cannot catch a glimpse of these occluded parts of reality, then neither can any other way of knowing. No revelation, *sensus divinitatis*, personal intuition, or inner sense of certainty will succeed where science has failed. Squaring the circle is not a "limit to geometry," as if there were some other magical way to use a compass and a ruler to pull off the trick. This, however, is exactly what is proclaimed by many defenders of methodological naturalism as a limit to science. Many theologians will gravely pronounce that the supernatural realm is beyond the limits of science, then jump quickly to the conclusion that their favorite way of knowing can lift the veil of mystery. As the theologian Paul de Vries, possibly the originator of the principle of methodological naturalism, put it:

> If we are free to let the natural sciences be limited to their perspectives under the guidance of methodological naturalism, then other sources of truth will become more defensible. However, to insist that God-talk be included in the natural sciences is to submit unwisely to the modern myth of scientism: the myth that all truth is scientific. (de Vries 1986, 396)

The theological agenda behind methodological naturalism and the invoca-
tion of scientism is clear. Science must be curtailed and reined in, so that
somehow, by default, room for other ways of knowing will become vacant.
That room can then be swiftly occupied by religious revelation, spiritual experience, or the deliverances of our *sensus divinitatis*, all of which are much
more profound than the vulgar methods of empirical inquiry. Under such an
arrangement, concludes the theologian John Haught, "theology is . . . freed
from moonlighting in the explanatory domain that science now occupies, so
that it may now gravitate toward its more natural setting—at levels of depth
to which science cannot reach" (2004, 236).

This use of the scientism gambit as an immunizing strategy for protecting
the supernatural realm from scientific incursions often occurs in apologist
literature (see Law, this volume). In a recent concerted attack on scientism
published with Bloomsbury (Williams and Robinson 2014), the editors write
that the most worrying aspect of scientism is its "insistence on naturalist, materialist, metaphysical orthodoxy" (Williams 2014, 7). Within the worldview
of scientism, they lament, there is no longer room for any sort of "transcendence," the "essence of our human relatedness" (8), and "trans-situational
values and morality generally" (17). In the closing lines of this sermon against
the evils of scientism, the reader is enjoined to go forth and "bear witness
to transcendence," so that those lost souls still under scientism's spell may
begin to "hear a still voice that whispers beyond the molecules, inviting us
to so much more" (188). Can the term *scientism* still be redeemed after such
ramblings?

The Moral Realm

Is there any instance, then, of science genuinely overreaching, pretending to
offer a form of knowledge that is in principle inaccessible, no matter how
broadly you construe science?[14] If any candidate fits the job description of the
concept of scientism, I think it is the notion, promoted by authors such as Sam
Harris (2011) and Richard Carrier (2011), that science can provide an objective
basis for morality. Moral values, these authors claim, are objects of scientific
investigation, just like electrons, DNA, and inflation. However, I think this
issue was settled almost three centuries ago by David Hume (1739), when he
noted that it is logically impossible to derive *ought* from *is*. Many have tried
to cross the is/ought gap, but to no avail. Moral realism is as baseless as it was
in Hume's day, and scientific advances since Hume's day will offer little help

(for a recent critical assessment, see Blackford 2016, chap. 4). What invariably happens in arguments for moral realism is that some *ought* premise is surreptitiously smuggled in among the *is* premises. Harris, for instance, already assumes the value of the well-being of conscious creatures from the start, only to then see it "confirmed" in scientific research (Blackford 2010). But that's like "finding" the Easter egg that you've just hidden away yourself. While I, too, along with Harris, would vote for a criterion like "the well-being of conscious creatures" as a starting point for ethical reasoning, it makes no sense to argue that this value is *itself* a finding of science. Science is invaluable in resolving moral disputes, but only after some moral value has been plugged in.[15]

If Hume is right, "ought" cannot be derived from "is" *simpliciter*, regardless of whether the "is" comprises scientific knowledge or any other sort of knowledge.[16] In this context, at least, it does not matter how broad and inclusive your definition of science might be. Even if you bring the social sciences and the humanities to the table, you will not make any headway in establishing objective moral facts. No method continuous with the sciences, in its broadest possible sense, can offer any more hope in ferreting out moral facts. This is because there aren't any (Mackie 1977). Which means that again, the problem is not so much a limitation of science per se: if moral facts do not exist, it is hardly surprising that no method is capable of finding them.[17]

Still, even if we admit that no other way of knowing will succeed where science fails, Harris and Carrier still seem to have a strangely misplaced confidence in the specialized sciences. Even if you are a moral realist and you think that Hume got it all wrong, it is unclear why you would need *neuroscience* or other specialized sciences to establish moral facts. If there are such queer entities as "moral facts" out there in the world, why expect to find them under a brain scanner? Harris, in particular, imagines that sophisticated equipment (fMRI scans and the like) will somehow carry us across a barrier that would otherwise be unbridgeable, objectively validating the well-being and suffering of conscious creatures. But why would we need to put someone under a brain scanner to find out, say, that he is in pain? Wouldn't it suffice to hear howls of agony (Pigliucci 2011)?[18]

In any case, if Hume was right, neither everyday knowledge nor cutting-edge science will bring us one step closer to bridging the gap. And if Mackie is right, there simply are no moral facts. Giving a bizarre doctrine like moral realism a patina of neuroscientific legitimacy won't make the position any more coherent. Perhaps this misplaced confidence in the deliverances of (neuro)science qualifies as a form of scientism.[19]

Conclusion

Does science have limits? Sure, the concept of science has some (fuzzy) practical, institutional, and semantic boundaries, but these are not relevant to the notion of *epistemic* limits, and the question of other ways of knowing besides science. For a naturalist, science is continuous with everyday knowledge, and enmeshed with other academic endeavors. Science as we know it results from the interplay between our cognitive makeup and the world. Our minds and the world out there are like the two blades of a pair of scissors, to use Herbert Simon's (1955) memorable metaphor. The alignment of the two blades is what makes science work. But in a different world, or with a different brain, science would look quite different indeed.

Does this unified web of knowledge break down somewhere? I have tried to find a domain of reality that would call for a genuinely different way of knowing sufficiently discontinuous with the sciences to deserve a name of its own. The supernatural realm won't do. As long as supernatural entities make *some* contact with the natural world, leaving *some* traces, science has a foothold to investigate them. The idea that science is by its very nature restricted to natural causes and explanations does not hold up to scrutiny. Science is what has worked in the past. The conspicuous absence of deities from science is a contingent outcome, not an a priori limitation to its methods. The doctrine of methodological naturalism as an intrinsic limitation of science is an immunizing strategy devised by theologians, and abetted by "believers in belief" (Dennett 2006), to protect supernatural doctrines from scientific scrutiny.

Science could smoke out ghosts and gods, if they existed, but it can garner no factual knowledge about the moral realm, for the simple reason that the notion of a moral fact is incoherent. No intuitive faculty or revelation or mystical procedure will bridge the is/ought gap. Here, perhaps, we have a suitable referent for the term *scientism*. By claiming that (neuro)science can bridge the is/ought gap and objectively ground morality, Harris and Carrier seem to demand the impossible from science: solving a type of problem that, in fact, cannot be solved by any method of inquiry at all.

Some may argue that the brand of holistic naturalism defended in this chapter is itself a form of scientism (Sorell 2002), as it appropriates all ways of knowing to science in the broad sense, leaves no room for religion or objective morality, and conceives of science as an infinitely flexible and open-ended endeavor without fixed methods of rules. To end this chapter, I will perhaps confirm their worst suspicions. If a factual question is answerable at all, it can be answered using methods that are at least *continuous* with science. If some epistemic enterprise becomes too detached from science, and thus from the rest of the web of knowledge with which science is connected, that usually does not bode well for that enterprise (e.g. theology, analytic metaphysics, phenomenology).

The fact that we have a convenient word for our systematic investigations into the natural world (*science*) has been an enormous source of confusion. It has often invited people to think that science depends on something called the scientific method, a simple procedure or formula that will invariably yield fruit when appropriately applied. Following down that road, it is natural to think of this scientific method as just one tool in our toolkit, next to other tools, each with its own domain of application. But it is a mistake to think that the word *science* picks out a clearly defined set of methods and procedures, which can be separated from other tools and methods. And if we cannot define science, then the problem of scientism evaporates.

In the end, the family of spatial metaphors that conceive of science as operating in a certain "realm," with identifiable "limits" or "borders," is misleading. Science is not like a ship equipped for navigating some waters but not others, and reality is not like a geography map with strange, uncharted regions that are accessible only through some "other way of sailing." If there are any limits to human knowledge, they will coincide with the limits of science, broadly construed. Perhaps someday the scientific enterprise will fall off the Edge of Knowledge, or slam up against a wall of blank incomprehension; but when that day arrives, no other way of knowing will be of any avail.

Acknowledgments

Many thanks to Russell Blackford, Krzysztof Dołęga, Dan Dennett, Lieven Pauwels, Massimo Pigliucci, Tom Sorell, Taner Edis, Leonardo Ambasciano, Filip Buekens, and several commentators on Academia.edu for stimulating discussions and helpful feedback. Part of the inspiration for this chapter arose from "The Limits of Science," a debate with Daniel Dennett, Lawrence Krauss, and Massimo Pigliucci, which I was asked to moderate for *Het Denkgelag* (YouTube video available at https://www.youtube.com/watch?v= 9tH3AnYyAI).

······

Notes to Chapter Two

1. In fact, it was the frontispiece to his *Instauratio magna*, the much larger unfinished work of which *Novum Organum* was just a part.

2. As Haack justifies the pejorative meaning: "As the English word 'scientism' is currently used, it is a trivial verbal truth that scientism . . . should be avoided. But it is a substantial question when, and why, deference to the sciences is inappropriate or exaggerated" (2012, 75).

3. I use *naturalism* here in the Quinean sense of holism about knowledge, not in the sense of a rejection of the supernatural. For a discussion of both senses, see Boudry (in press).

4. Centaurians would probably have their own innate folk knowledge of one another's mental architecture (if they are a social species).

5. The plumbing-is-science trope is due to the biologist Jerry Coyne, and created something of a stir on various blogs and websites (Coyne 2015, 40–41).

6. "Inuit Hunters Help Scientists Track Narwhals," *NPR*, August 19, 2009, http://www.npr.org/2009/08/19/111980557/inuit-hunters-help-scientists-track-narwhals.

7. This is not strictly true. In many scientific areas, it takes a well-trained expert to spot the relevant patterns. A paleontologist can "see" a fossil where a layperson just sees dust and pebbles. But the point still stands that all science starts and ends with perception.

8. If preoccupation with this normative demarcation is itself a major symptom of the syndrome of scientism, as Haack (2012) claims, then I gleefully accept the diagnosis.

9. People in the humanities and social sciences are often confronted with the charge that they do not live up to some outdated ideal of science (for instance that science should be law-like, mathematical, predictive, employ statistical methods, etc.). In response to such objections, some have given up the pretension to be scientific altogether, and have adopted a different standard, characterizing their discipline as an art or craft instead. Others have tried to emulate this outdated ideal of science, unaware that not even modern physics adheres to it. Is this a form of scientism, then? Some critics have thought so, but the problem with this usage is that it actually *reinforces* those same misconceptions about the nature of science. Instead of arguing that people overextend the methods of science in the humanities, we should update their understanding of science, and show them that not even physics answers to their ideal of what science should be. See Ross, this volume, who makes this argument in the context of economics.

10. Even a deist believes that God has acted as a primordial cause of the universe, somehow lighting the fuse of the big bang, and withdrawing from the resulting mess thereafter.

11. Or perhaps they can materialize selectively, when they desire to? But then aren't they violating the laws of thermodynamics? Ghostologists have some work to do. Thanks to Taner Edis for bringing up this point.

12. There are, however, stable empirical generalizations in the social sciences.

Although these are sometimes referred to as laws, they are nothing like the nomological laws of historicism. Thanks to Lieven Pauwels for this point.

13. It will not do to argue that one needs philosophy to rule out such claims, and that science *by itself* cannot rule out Last Thursdayism. Ockham's razor is as much a tool in the scientist's toolkit as it is in the philosopher's. This shows again that it is fruitless to pry apart philosophy from science.

14. Some people might say that science forms a clear borderline with pure mathematics and logic, but it is debatable whether mathematics offers any genuine knowledge. I follow Kitcher (2012) in thinking of mathematics as a set of elaborate games with certain rules and moves, where some games are more interesting than others, and "for which ascriptions of truth are used to mark out the results of preferred moves" (Kitcher 2012, 185). In any case, many branches of science are heavily dependent on mathematics, and few would argue that including mathematics under the umbrella of science is a form of scientism.

15. This, of course, is no excuse for moral philosophers to be complacently ignorant about the empirical findings of psychology, neurology, and evolutionary science relating to morality. For a nuanced plea for an empirically informed moral philosophy, which still respects a "sharp and crucial distinction" between is and ought, see Greene (2003).

16. Ironically, Williams (2014), along with many others, takes scientism to entail the *denial* of moral realism. So which is it: are moral realists (of Harris's [2011] stripe) the scientistic ones, or rather those who reject Williams's views? Perhaps this shows that the term *scientism* is just not the most useful term in these discussions.

17. In this respect, G. E. Moore's "naturalistic fallacy" is a misnomer, because it suggests that even though there is no way to define what is moral in natural terms, there may be some other, nonnatural way to ground our conception of what is morally good.

18. There is interesting scientific (!) research on why neurological explanations have a special allure, even when such explanations are redundant: see Weisberg et al. 2008. Thanks to Filip Buekens for this reference.

19. Despite all of this, I actually sympathize with Harris's inclusive definition of science, as in his response to the 2014 Edge.org question: "Our Narrow Definition of 'Science,'" https://www.edge.org/response-detail/25372. Accessed on February 14, 2017.

Literature Cited

Austin, J. L. 1962. *How to Do Things with Words.* Cambridge, MA: Harvard University Press.

Blackford, Russell. 2010. "Book Review: Sam Harris' *The Moral Landscape.*" *Journal of Evolution and Technology* 21, no. 2 (December): 53–62.

———. 2016. *The Mystery of Moral Authority.* Basingstoke: Palgrave Macmillan.

Boudry, Maarten. 2013. "Loki's Wager and Laudan's Error: On Genuine and Territorial Demarcation." In *Philosophy of Pseudoscience: Reconsidering the Demarcation Problem*, edited by M. Pigliucci and M. Boudry, 79–98. Chicago: University of Chicago Press.

———. 2015. "The Sin of Scientism: Response to Clark." *Reports of the National Center for Science Education* 35 (5): 61–62.

———. In press. "A Most Unnatural Alliance." *Free Inquiry.*

Boudry, Maarten, Stefaan Blancke, and Johan Braeckman. 2010. "How Not to Attack Intelligent Design Creationism: Philosophical Misconceptions about Methodological Naturalism." *Foundations of Science* 15 (3): 227–44. doi: 10.1007/s10699-010-9178-7.

———. 2012. "Grist to the Mill of Anti-evolutionism: The Failed Strategy of Ruling the Supernatural Out of Science by Philosophical Fiat." *Science and Education* 21 (8): 1151–1165. doi: 10.1007/s11191-012-9446-8.

Bynum, W. F., E. J. Browne, and R. Porter. 2014. *Dictionary of the History of Science.* Princeton, NJ: Princeton University Press.

Carrier, Richard. 2011. "Moral Facts Naturally Exist (and Science Could Find Them)." In *The End of Christianity*, edited by John W. Loftus, 333–64. Amherst, NY: Prometheus Books.

Carroll, Sean. 2016. *The Big Picture: On the Origins of Life, Meaning, and the Universe Itself.* New York: Dutton, an imprint of Penguin Random House.

Clark, Kelly James. 2015. "Is Theism a Scientific Hypothesis? Reply to Boudry." *Reports of the National Center for Science Education* 35 (5): 51–53.

Cleland, Carol. E. 2002. "Methodological and Epistemic Differences between Historical Science and Experimental Science." *Philosophy of Science* 69 (3): 474–96.

Coyne, Jerry A. 2015. *Faith versus Fact: Why Science and Religion Are Incompatible.* New York: Viking.

Dennett, Daniel C. 1981. "True Believers: The Intentional Strategy and Why It Works." In *Scientific Explanation: Papers Based on Herbert Spencer Lectures Given in the University of Oxford*, edited by A. F. Heath, 150–67. Oxford: Clarendon.

———. 1993. *Consciousness Explained.* London: Penguin.

———. 1995. *Darwin's Dangerous Idea: Evolution and the Meanings of Life.* New York: Simon and Schuster.

———. 2006. *Breaking the Spell: Religion as a Natural Phenomenon.* New York: Penguin Books.

Edis, Taner. 2002. *The Ghost in the Universe: God in Light of Modern Science.* Amherst, NY: Prometheus Books.

Fishman, Yonatan I., and Maarten Boudry. 2013. "Does Science Presuppose Naturalism (or Anything at All)?" *Science and Education* 22 (5): 921–49.

Fodor, Jerry. 1983. *The Modularity of Mind: An Essay on Faculty Psychology.* Cambridge, MA: MIT Press.

Forrest, Barbara. 2000. "Methodological Naturalism and Philosophical Naturalism: Clarifying the Connection." *Philo* 3 (2): 7–29.

Greene, Joshua. 2003. "From Neural 'Is' to Moral 'Ought': What Are the Moral Implications of Neuroscientific Moral Psychology?" *Nature Reviews Neuroscience* 4 (10): 846–50.

Grice, Paul. 1989. *Studies in the Way of Words.* Cambridge, MA: Harvard University Press.

Haack, Susan. 2007. *Defending Science—within Reason: Between Scientism and Cynicism.* Amherst, NY: Prometheus Books.

———. 2012. "Six Signs of Scientism." *Logos and Episteme* 3 (1): 75–95.

Harris, Sam. 2011. *The Moral Landscape: How Science Can Determine Human Values*: New York: Simon and Schuster.

Haught, John F. 2004. "Darwin, Design, and Divine Providence." In *Debating Design: From Darwin to DNA*, edited by M. Ruse and W. A. Dembski, 229–45. Cambridge: Cambridge University Press.

Hull, David L. 1990. *Science as a Process: An Evolutionary Account of the Social and Conceptual Development of Science.* Chicago: University of Chicago Press.

Hume, David. 1739. *A Treatise of Human Nature.* Gutenberg Project, https://www.gutenberg.org/files/4705/4705-h/4705-h.htm.

Huxley, Thomas Henry. 1871. *Mr. Darwin's Critics.* London: Isbister. http://fullonlinebook.com/download-book/book/itoz/mr-darwin-s-critics.pdf.

Johnson, Phillip E. 2002. *The Wedge of Truth: Splitting the Foundations of Naturalism.* Downers Grove, IL: IVP Books, an imprint of InterVarsity Press.

Kitcher, Philip. 2012. *Preludes to Pragmatism: Toward a Reconstruction of Philosophy.* New York: Oxford University Press.

Ladyman, James, and Don Ross. 2007. *Every Thing Must Go: Metaphysics Naturalized.* Oxford: Oxford University Press.

Law, Stephen. 2011. *Believing Bullshit: How Not to Get Sucked into an Intellectual Black Hole.* Amherst, NY: Prometheus Books.

Mackie, John. 1977. *Ethics: Inventing Right and Wrong.* London: Penguin Books.

Pennock, Robert T. 1999. *Tower of Babel: The Evidence against the New Creationism.* Cambridge, MA: Bradford Books, an imprint of MIT Press.

Pigliucci, Massimo. 2011. "Science and the Is/Ought Problems: A Review of 'The Moral Landscape: How Science Can Determine Human Values' by Sam Harris." *Skeptic* 16 (3): 60–62.

Pigliucci, Massimo, and Maarten Boudry, eds. 2013. *Philosophy of Pseudoscience: Reconsidering the Demarcation Problem.* Chicago: University of Chicago Press.

Pinker, Steven. 2013. "Science Is Not Your Enemy." *New Republic*, August 6, https://newrepublic.com/article/114127/science-not-enemy-humanities.

Popper, Karl Raimund. 1957. *The Poverty of Historicism.* London: Routledge and Kegan Paul.

Putnam, Hilary. 2013. *Meaning and the Moral Sciences.* London: New York: Routledge Revivals, an imprint of Routledge Taylor and Francis Group.

Quine, Willard Van Orman. 1951. "Two Dogmas of Empiricism." *Philosophical Review* 60: 20–43.

Rosenberg, Alex. 2011. *The Atheist's Guide to Reality: Enjoying Life without Illusions.* New York: W. W. Norton.

Russell, Bertrand. (1921) 2005. *The Analysis of Mind.* London: Allen and Unwin.

Sellars, Wilfred. 1963. "Philosophy and the Scientific Image of Man." In *Science, Perception and Reality*, 1–40. London: Routledge and Kegan Paul.

Simon, Herbert A. 1955. "A Behavioral Model of Rational Choice." *Quarterly Journal of Economics* 69 (1): 99–118.

Sorell, Tom. 2002. *Scientism: Philosophy and the Infatuation with Science*. London: Routledge.

Sterelny, Kim. 2010. "Minds: Extended or Scaffolded?" *Phenomenology and the Cognitive Sciences* 9 (4): 465–81.

Tavris, Carol. 1988. "Beyond Cartoon Killings: Comments on Two Overlooked Effects of Television." In *Television as a Social Issue*, edited by S. Oskamp, 189–97. Newbury Park, CA: Sage.

Thagard, Paul. 2012. *The Cognitive Science of Science: Explanation, Discovery, and Conceptual Change*. Cambridge, MA: MIT Press.

Vries, Paul de. 1986. "Naturalism in the Natural Sciences: A Christian Perspective." *Christian Scholar's Review* 15 (4): 388–96.

Weisberg, Deena Skolnick, Frank C. Keil, Joshua Goodstein, Elizabeth Rawson, and Jeremy R. Gray. 2008. "The Seductive Allure of Neuroscience Explanations." *Journal of Cognitive Neuroscience* 20 (3): 470–77. doi: 10.1162/jocn.2008.20040.

Wieseltier, Leon. 2013. "Crimes against Humanities: Now Science Wants to Invade the Liberal Arts; Don't Let It Happen." *New Republic*, September 3, https://newrepublic.com/article/114548/leon-wieseltier-responds-steven-pinkers-scientism.

Williams, Richard N. 2014. Introduction to *Scientism: The New Orthodoxy*, edited by Richard N. Williams and Daniel N. Robinson, 1–21. London: Bloomsbury Academic.

Williams, Richard N., and Daniel N Robinson, eds. 2014. *Scientism: The New Orthodoxy*. London: Bloomsbury Academic.

3

Scientism and the Argument from Supervenience of the Mental on the Physical

FILIP BUEKENS

"Physical Facts Fix All the Facts"

The idea that everything supervenes on the physical, or that the physical facts fix all the (experiential, mental, social . . .) facts, figures prominently in a broad family of arguments that appeal to those who have antecedently convinced themselves that physics, the science which *by definition* explores the physical and nothing but the physical, explains everything. Alexander Rosenberg makes no secret about the aims and claims of this line of argument:

> What is the world really like? It's fermions and bosons, and everything that can be made up of them, and nothing that can't be made up of them. All the facts about fermions and bosons determine or "fix" all the other facts about reality and what exists in this universe or any other if, as physics may end up showing, there are other ones. Another way of expressing this fact-fixing by physics is to say that all the other facts—the chemical, biological, psychological, social, economic, political, cultural facts—supervene on the physical facts and are ultimately *explained* by them. And if *physics* can't in principle fix a putative fact, it is no fact after all. (Rosenberg 2015; my italics)[1]

Many contemporary philosophers have subscribed to the supervenience of the psychological on the physical, thus joining Rosenberg in his rejection of dualism as a remnant of cognitive biases and unfounded commonsense intuitions. But one of the most salient and disturbing consequences of his particular brand of scientism would be the gratuitousness of our folk psychological concepts and the explanations in which they figure, including for example the concept of truth, and explanations of the importance of having true beliefs (see also Rosenberg's contribution to this volume). Rosenberg endorses supervenience of the mental on the physical (the substance) in the sense that the physical facts fix all the facts, but he goes on, in the penultimate sentence of this quote, to say that all the other facts are also ultimately *explained* by physics (the science)—and that is, as I shall argue in this chapter, a much stronger claim. There is a version of supervenience on the philosophical market that endorses the claim that the physical facts fix all the facts, but holds that this is trivially true. Why? Because physicalists must hold that a complete physics (the science, including the laws and models it proposes) can give us descriptions that individuate states, objects, and events: when two states, objects, or events fall under the same individuating physical description, they are identical, hence share all the other ("supervenient") properties. This is, as I shall explain, consistent with conceptual antireductionism. Conceptual antireductionism holds that explanations employing psychological concepts cannot be replaced by explanatory strategies relying on physical concepts and physical evidence. If it is true (in fact, trivially true) that the physical facts in a world w fix all the other facts in w, the brand of scientism Rosenberg endorses owes us a further argument for explanatory reductionism. My aim in this chapter is not to provide such an argument against explanatory reductionism but, more modestly, to explore the aims and claims of supervenience as a legitimate concept in the hands of nonreductive monists, and to explain how it can be true, according to nonreductive monists, that physical facts fix all the facts.

Let me begin by sketching a plausible monistic conception of the world. It is a conception which holds that some physical objects, events, states, or processes are *identical* with mental or experiential phenomena (objects, events, or states). Every concrete, existing mental state or event (a state or event that falls under a true mental description) is numerically identical with something that falls under a true *physical* description. Identity itself is a perfectly clear extensional relation in nature, in fact the strongest there can possibly be. Every thing/state/object has it to itself, and nothing has it to something other than itself. This "ontological monism" (Donald Davidson [1980]) or "real materialism" (Galen Strawson [2008]) or "antecedent physicalism" (John Perry [2002]) is motivated by the manifest fact that any real, concrete mental state or event is part of the causal order, and causal *relata* also fall under true *physical* descriptions.

This brand of monism (sometimes inspired by Spinozistic monism) holds that "being mental" and "being physical" are qualifications of a process, event, state . . . relative to true mental or physical descriptions we can give of one and the same entity. There is, on this account, one substance, with different (biological, mental, social . . .) manifestations (see again Strawson 2008). The usual caveats about abstracta apply and won't be dealt with here. Even Rosenberg himself acknowledges that (knowledge of) mathematics is a problem (yet another!) for scientism.[2] Ontological monism is consistent with the nonreducibility of mental and psychological concepts, as well as the explanatory practices in which they figure, to the kind of concepts and explanatory practices that are part of our best physics. (The deeper problem is that the nature of physics must allow for mental facts. Does current physics itself grasp the full nature of the physical? Bertrand Russell, for one, had his doubts, and they are further explored by Strawson 2008).

Supervenience

There is a version of supervenience of the mental on the physical that is consistent with ontological monism and conceptual dualism, but it's not Rosenberg's. It is important to remind ourselves here that supervenience is a *technical* concept invented by philosophers, and that "fixing the facts" as used in "physical facts fix all the mental or psychological facts" is a technical locution in need of a nontrivial elucidation. Its elucidation must avoid bias created by excessive exposure to science ("nothing counts as knowledge unless it is scientific knowledge") or a reductionist or physicalist impulse driven by, for example, antireligious concerns. Moreover, it should also be noted that Rosenberg's reductive use of supervenience would have surprised, for example, G. E. Moore, who held that moral properties are not subject to naturalistic explanations, even though they supervene on natural properties. For reductionists, supervenience goes hand in hand with reductive explanations — witness again the move from "everything supervenes on the physical" (a claim about a substance) to "physics gives the ultimate explanation" (a claim about a web of theories and models).

Rosenberg needs a strong version of supervenience, and contemporary philosophers have given him duplicative accounts. A couple of decades ago, supervenience was standardly formulated as a claim about what follows concerning the further (supervening) properties of two entities when they are *physical duplicates*. Some examples suffice here. J. Owens holds that "it is necessarily true that if two worlds are alike in every physical respect, then they are exactly similar in every respect. . . . Once the physical features of a world are determined its non-physical features are determined too" (1989, 60). David Papineau held that "if two systems are physically identical, then they must also be chemically identical, biologically identical, psychologically identical, and so on" (1993, 10). David Lewis leaves no doubts as to what is at stake: "Interesting supervenience theses usually involve the notion of qualitative duplication . . . we must make sense of duplication even to ask the question [of supervenience]" (1983, 358). Similar claims can be found in many others defending versions of antidualism combined with determination-claims.

"Indiscernibility with respect to the subvenient class of properties," "physical indiscernibility," "being exactly alike in all natural respects," and "qualitative duplication" are the relevant notions here. The idea was that identity could be *relativized* to certain *relevant* properties, thus creating *types* of entities. The template for duplicative accounts of supervenience as applied to the mental/physical was something like this:

(Supervenience Template)
Necessarily, for any mental property M, if anything has M, there exists a (complex) physical property P such that it has P, and necessarily, anything that has P has M.

Variations typically concerned issues about whether the relevant physical properties should be confined to intrinsic properties of the brain or include relational properties, relating brain states/mental states to their environment or causal history. Externalists about mental properties—philosophers who thought, sensibly enough, that the individuation of a mental state or event required looking beyond the skull and skin of the agent—typically defended the inclusion of relational properties and strong modal versions of the template. The most popular version held that two worlds which are physically indiscernible are also psychologically indiscernible. The last conjunct of the Supervenience Template ("anything that has P has M") introduced the projection claim, and accounts for the *adagium* that physical facts fix or necessitate all the facts (including the mental or psychological facts).

But there was a subtle alternative on the market that was heavily criticized but seldom well understood. Donald Davidson's definition of supervenience, first formulated in his seminal paper "Mental Events" (1969), was *not* based on a duplication thesis. In that paper, which single-handedly reintroduced the concept in contemporary philosophy of mind, Davidson offered and endorsed a nonduplicative reading of supervenience as applied to the mental/physical:

(Davidson on Supervenience)

Mental characteristics are supervenient on physical characteristics. Such supervenience might be taken to mean that there cannot be two events alike in all physical respects but differing in some mental respect, or that an object cannot alter in some respect without altering in some physical respect. Supervenience of this kind doesn't entail reducibility through law or definition. (Davidson [1969] 1980, 214)

Davidson's definition does not rely on physical duplication. There is talk of two *particular* events (or *token* events), not *types*, and we are told that no two events can be alike in all physical respects (the individuating respects are part of the physical respects) yet differ in some other respect, which is trivially true since no two token events can be alike in all physical respects and yet be different events. In this second definition, there is talk of change in an object, and the idea is that any change entails a physical change. A similar proposal was voiced by W. V. Quine (in a formulation undoubtedly influenced by Davidson): "Nothing happens in the world, not the flutter of an eyelid, not the flicker of a thought, without some redistribution of microphysical states" (1981, 98).

It is this nonduplicative version of supervenience that I am going to explore in this chapter, and mobilize against the stronger version that figures in Rosenberg's brand of scientism. My claim is that nonduplicative accounts support the first claim Rosenberg makes—that physical facts determine all the facts—but are consistent with a rejection of the second claim—that physical facts do not explain all the facts. I will then show that the claim that "the physical facts fix all the facts" boils down to the trivial claim that, if two worlds w and w' are physically identical, and monism is true (there are no strange substances responsible for the mental facts), then those worlds are identical *tout court*, that is, identical in the Leibnizian sense. The totality of facts about w trivially fixes the facts about w'. The idea that physical facts fix all the facts is as trivial as the idea that all the facts about the morning star, or water, fix the facts about the evening star, or H_2O. This is hardly an interesting result, and one that can be fully endorsed by nonreductive physicalists.

For the conceptual antireductionist, the challenge is to specify what kind of dependency of the mental on the physical one could defend that is consistent with conceptual dualism. The latter position states that the way we conceptualize the realm of the mental is constitutively different from the way we conceptualize those very same states and events, but using concepts drawn from physics. Davidson's account of supervenience, combined with some independently plausible claims, did exactly that: it promised dependence (hence approves of physical facts as "fixing all the facts") but not explanatory reductionism.

First, I'll explore, following a formal template proposed by Davidson, supervenience as a relation between a predicate and a set of predicates ("D-supervenience"). Then I examine which further constraints must be added to D-supervenience at the ontological and ideological (i.e. the level of predicates involved) to arrive at a version that would support Rosenberg's reductive reading of supervenience. It will turn out that one of the constraints reductive physicalists must add to D-supervenience is the supposition that physical indiscernibility does *not* collapse into numerical or Leibnizian identity: individuating physical facts are excluded from the projection base that figures in the Supervenience Template. It will then be argued that physical indiscernibility, according to physicalism, *does* collapse into Leibnizian identity, which trivializes the "fixing the further facts"-claim. According to nonreductive physicalists, we can hold that the physical facts fix all the facts without having to endorse reductionism or eliminativism.

D-supervenience: An Uncontroversial Application

I will first look at a "dry" formulation of Davidson's concept of supervenience, basically a stipulated dependency-relation between sets of predicates ("D-supervenience"). Examining the possibilities and limits of a formal template independently of the kinds of discourse it applies to makes it clear which *additional* requirements must be fulfilled in order to strengthen it into a reductive relation useful in scientistic programs. Davidson's template ran as follows:

[D]

A predicate p is supervenient on a set of predicates S if for every pair of objects such that p is true of one and not of the other there is a predicate of S that is true of one and not of the other. (Davidson 1985, 242)

Consider a noncontroversial application of [D]. Let the supervening predicate *p* characterize an individual in terms of the country where (s)he resides, and the subvenient predicates of *S* characterize the same individual in terms of the city where (s)he lives. Assume, for the sake of the argument, that all relevant individuals in the domain to which [D] applies live in cities (this assumption will be dropped later). Trivially, if two individuals live in different countries they must live in different cities (assuming, for the sake of the argument, that no city is located in two different countries). By contraposition, if two subjects live in the same city (if they are *city-indiscernible*), they must live in the same country. No two subjects *x* and *y* who share city-characteristics differ in their country-characteristics. Put in terms of changes: if an individual moves from one country to another, her city-characteristics change (recall that it was assumed that all individuals in the domain of this application of [D] live in cities).

Since we can allocate every city to exactly one country, we can infer from the city in which an individual lives the country where the subject resides. City-characteristics and country-characteristics can therefore be connected via "city-country" bridge laws. An exhaustive list of bridge laws would be very interesting. It would show that when it comes to locating individuals, the city-predicates provide a more precise location of the individual. If our project is the location of individuals, predicates that characterize the subvenient basis are more accurate than the predicates that characterize the supervenient structure. We could dispense with the country-vocabulary, if we want, and, using the bridge laws, see country-characteristics as wholly dependent on the more accurate city-characteristics. (The increased accuracy of the city-vocabulary as compared with the country-vocabulary assumes that the projects in which both vocabularies function—locating individuals, for example—remain the same.)

The example also illustrates that supervenience leaves room for two persons *x* and *y* to share the same country-predicate without having the same city-characteristics. Let us say that if this is the case, the city-vocabulary—the subvenient vocabulary—is *not closed*: it does not apply to everything in the domain of the country-vocabulary (i.e., those who live in small villages). And due to *lack of expressive power*, the city-vocabulary does not allow us to single out an individual *uniquely* via her city-characteristics: different persons can share the same city-characteristic—they are, with respect to that city-characteristic, *type-identical*. All we can infer, using the city-country bridge laws, is that members of the set of individuals who are "city-identical" (who are of the same city-type) live in the same country.

In this example, and now returning to [D], the p-characteristic or property can be defined or even reduced away in favor of characteristics in S. The fact that we can define "being located in country k" (with variable k) in terms of "being located in a city c in country k" illustrates that D-supervenience doesn't *formally exclude* that the p-predicate can be defined in terms of S-predicates, that is, that the supervenient p-predicate can itself be a member of S. That p does not belong to S (if the constitutive principles that govern the application of p are not part of those that govern the application of members of S) is an *additional* antireductionist claim about predicates or concepts, consistent with, but not entailed by, the supervenience of p-facts on S-facts. Antireductionism with respect to p relative to S is a further substantial claim about the characteristics related, not part of the template provided by D-supervenience. Antireductionists will, according to this model, challenge the idea that the mental or psychological explanations in which mental concepts or characteristics figure can be reduced to or help resolve explanatory concerns served by physical predicates or characteristics. And one of their key arguments holds that explanatory projects in which the mental concepts play a key explanatory role answer to different constitutive principles than those governing members of S—the physical predicates. (These points are well known, and versions of the argument are defended in Davidson 1980; McDowell 1986; Child 1994.) Many have challenged this family of arguments, not in the least proponents of scientism. But that is not key to my argument. My key point will be that a Davidsonian can hold that the physical facts fix all the other facts, and that this is consistent with antireductionism.

The example shows why D-supervenience implies that the ontology of the country-vocabulary need not be richer than the ontology of the city-vocabulary. The p-predicate is true of the same entity the S-predicates are true of. More important, city/country-supervenience doesn't entail that the city-vocabulary contains predicates that are *uniquely true* of an individual, in other words, that individuate him or her. This reflects an important sense in which the city-vocabulary differs from what an ideal physics should offer: the fact that everything a mental predicate is true of has *unique physical characteristics* is, as we shall see, a feature of the expressivity of the system of physical predicates. D-supervenience itself is silent about the expressive power of the predicates in S.

Finally, D-supervenience also remains silent with respect to the kinds of explanations and explanatory projects that employ the predicates involved. It doesn't *entail* that the explanatory interests in which the use of predicate p

is mandatory must be the same as projects in which predicates drawn from the set S are relevant. The city-country example illustrates that if we are interested in finding out where x lives, S-predicates, that is, true descriptions in terms of S-predicates, will actually be a lot more informative than true descriptions employing p-predicates.[3] But this feature depends on the reducibility of p-predicates and the fact that, in this particular case, the explanatory projects in which S-predicates figure can fully absorb the p-predicates. Insofar as cities don't move from country to country (the "bridge laws" are assumed to be constant), and in view of the explanatory project in which the predicates figure, the subvenient city-predicates can subsume the supervenient country-predicates.

We now have enough material on board to apply D-supervenience to supervenience of the mental on the physical. Let m be a mental predicate, and S_p be a set of physical predicates (which belong to what scientism would call a "final physics," and assume for the sake of this chapter that there are no deep epistemic constraints on arriving at such a theory). We can then hold the following:

[MP]

A mental predicate m is supervenient on a set of predicates SP if for every pair of events such that m is true of one and not of the other there is a predicate of SP that is true of one and not of the other.

I will show that *if* the further conditions (i)–(iii), which are constraints on the predicates reflected by the subject matter of [MP], are satisfied, a form of nonreductive monism holds.

   (i) m is not a member of S_p (m is *not reducible* to any physical predicate—or complex combination of predicates—in S)
   (ii)   Every entity that falls under m has S_p-characteristics (the S_p-vocabulary is *closed*) and
   (iii)   If two entities e and e' share all their intrinsic and relational physical properties, they will be identical *tout court* (the vocabulary of S_p *individuates* events).[4]

Recall that the absence of analogous constraints on country/city-supervenience didn't render it inconsistent. Since (i)–(iii) do not follow from [MP], they are *independent* constraints on the predicates [MP] is applied to.

The first constraint—conceptual antireductionism—is not my concern here. My argument would broadly follow the well-established Davidson-McDowell line of argument, which stresses the irreducibly normative and anomalous nature of mental predicates (relative to physical predicates).[5] (If we don't accept the irreducibility of the mental—if we think, for example, that the mental can be reductively explained in terms of the physical—then what follows should be read as an important *conditional* thesis: there is no formal or conceptual inconsistency involved in accepting (i) + (ii).) Claim (ii) suffices for *monism* in the following ontologically shallow sense: if event *e* falls under a mental description, then it *eo ipso* falls under a physical description true of it; hence it is numerically identical with an event that falls under a physical description true of it. Davidson argued for this claim on the grounds that what is involved in causal relations must fall under descriptions drawn from lawlike generalizations, and those generalizations are to be found in physics (the science) (Davidson 1980). Alternatively, we might argue that the idea that every mental event or state also falls under a physical description is the consequence of a heuristic identity thesis. For all we know, if we encounter a mental state or event, we will eventually be able to give a true physical description of it, based on the rather obvious fact that any such state or event has causes and effects, hence must be inscribed in the physical order. This brand of monism sees no *tension* between "the mental" and "the physical." Some physical states or events are mental states or events, just as some animals are feathered animals.

Claim (iii) is going to be crucial in what follows. Physicalism, as I take it, claims that a complete physics will be able to produce physical descriptions which *uniquely individuate a world in which a state or event occurs.* If two worlds w_i and w_j share the same unique physical characteristics, then w_i and w_j are identical. Just as two sets are identical when their members are identical, so worlds are identical when they have the same unique characteristics. This excludes, as materialism and nonreductive physicalism requires, that there are nonphysical difference-makers between w_i and w_j. If so, talk about "physical facts fixing all the facts" is reduced to the triviality that, if two states or events share all their intrinsic and relational physical properties, they must be numerically identical, and an application of Leibniz's law of the indiscernibility of identicals trivializes "fixing the further facts"-rhetoric. No proponent of scientism should be tempted to avail herself at this point of a metaphysical conception of worlds based on Lewisian possibilism, in which individuals have "physically indiscernible counterparts" (Lewis 1986). If anything amounts to metaphysical extravaganza, then surely Lewisian worlds would be a prime example!

Physical Indiscernibility as a Projection Base

Duplicative accounts of supervenience hold that physics is capable of producing a (complex) physical characterization of an individual or an individual located in its world, such that any *other* individual (or individual in its world) which is physically indiscernible from it (i.e., with which it is type-identical) will have the same *nonphysical* characteristics. Physical indiscernibility entails biological, mental, social . . . indiscernibility, or, in Rosenberg's idiom, physical facts fix all the facts. My point will be that these "fixing the other facts"-claims are a trivial consequence of the third constraint, that a complete physics can individuate the entities mental descriptions are true of.

To see how duplicative accounts of supervenience can be put to work, consider how city/country-supervenience could be reformulated as an *indiscernibility* thesis. The standard move would be something like this: given the city-vocabulary, apply Quine's "maxim of the identification of indiscernibles": "objects indistinguishable from one another within the terms of a given discourse should be construed as identical for that discourse" (1953, 71). The Quinean approach introduces indiscernibility *relative to a discourse*. "The references to the original entities (different entities, of course) should be reconstructed (for the purposes of this paper: the physical discourse) as referring to other and fewer objects, in such a way that indistinguishable originals give way each to the same new object" (ibid., slightly adapted). The newly created object, an equivalence class of objects, is a *type*. The original entities, distinguishable in other discourses, are now taken to be identical.[6] It is crucial to see that this introduction of types is the result of *stipulation* rather than *inquiry*: in view of the fact that John and Mary (and many others) live in London, we *create* the type "(being a) Londoner" (the predicate "is London-identical"), which might or might not be useful for further classificatory purposes ("if x lives in London, then x lives in the UK," for example).

We seem to proceed in a similar fashion when "physically indiscernible entities" are introduced. A closer look at various versions of duplicative supervenience reveals why: the physically identical types must either be comparable physically identical individuals *in* a world (weak supervenience), or allow for *projecting* the nonphysical characteristics of an individual (in its world) onto other, counterfactually existing individuals which are physically identical with the first.[7] "If an x that is F also has M, then any other x with the same F will also have M." Both types of supervenience must make empirically informed claims about *which* intrinsic and relational physical characteristics are going to be relevant for determining when two individuals are going to be relevantly indiscernible in some further respect. When creating types, we must therefore first determine what the *relevant* physical characteristics are going to be if we intend to project "nonphysical" properties onto objects or individuals physically identical to the first. Creating physically indiscernible objects that are not numerically identical therefore involves important further empirical assumptions concerning what exactly is going to be physically responsible for a subject's being in a mental state M, such that any other entity with those physical features F has the M-features. Call the assumed physical features the projection base. Weak and strong supervenience make importantly different claims about scope or *range* of the projection base.

Local intraworld supervenience (weak supervenience) holds that the *comparison or projection base*—the range r within which physical indiscernibility must be secured—is confined to an individual's intrinsic physical properties. *Global* supervenience claims that the comparison or projection base is broader: the relational character of the psychological requires that the world in which the individual resides is going to be relevant as offering the subvening physical base. In that case, a projection claim from one world to another is required: global supervenience entails strong supervenience.

As I said, creating physically indiscernible objects—a type of object—that fall under a comparison- or projection-based supervenience claim requires further empirical assumptions about the nature of mental, contentful states which cannot simply be stipulated: which physical aspects of the world (including the physical aspects of the individuals in that world) are mentally relevant? *Localists* claim that the relevant physical type confines the range of the projection base to physical characteristics of the body of the individual. Tyler Burge, for example, has local physical characteristics in mind when he writes that "certain intentional mental properties of an individual [Burge refers to sensations, *FB*] are locally supervenient," which is to say that "they could not be different from what they are, given the individual's physical, chemical, neural or functional histories where these histories are specified non-intentionally in a way that is independent of physical or social conditions outside the individual's body" (1986, 4).

The physically relevant range we want to demarcate thus depends on our assumptions concerning the physical facts a true mental description depends on. If we think that for experiences only intrinsic physical features matter, we go for local, intrinsic physical facts. Depending on what kind of mental phenomenon was discussed, a wide variety of options concerning the range of the projection base has been proposed: the histories of the individuals must be physically identical, the natural kinds in the world must be identical (Putnam 1975), the worlds *tout court* must be identical (Grimes 1991).[8]

Let us first explore this issue for weak supervenience. The fact that two individuals are physically indiscernible but still comparable (they resemble each other in certain respects, not in others) means that there will be physical differences between them (their different spatiotemporal location, for instance, the fact that they are built out of different sets of atoms, that they have different physical histories, etc.). But weak supervenience, it is often argued, entails the epiphenomenal character of the mental properties so defined. The relevant psychological differences can be grounded outside the range r marked out by the projection base. Once we've accepted that two physically identical individuals (given the physical projection base r) can be psychologically different (that something outside the region r can make a psychologically relevant difference), we are no longer saying that the physical facts within the range r fix all the psychological facts about an individual.

Strong supervenience solves this by locating physically indiscernible individuals in physically indiscernible worlds. The relevant range r that provides the projection base is now a whole world, so no epiphenomenalism can follow. Physically identical individuals e and e' have that property by virtue of being located in physically identical worlds. Strong supervenience claims that if we have a world w_j which is physically identical with a world w_i in which a mental entity e resides, then e' in w_j will have exactly the same (mental, nonphysical) properties as e in w_i because the former world will be physically indiscernible from the latter.

Strong supervenience embodies the claims about "fixing the further facts" that the version of scientism discussed at the beginning of this chapter seems to require—a modal claim which allows us to *infer* mental characteristics from physical characteristics. Given a set of physically indiscernible worlds W, we can fix all the nonphysical characteristics of members of W if we know the nonphysical characteristics of one of its members. Given a complete physical description of one of the members of W, we can *predict* mental states of creatures in other members of W. The "projection" claim included in strong supervenience has the following format:

$$(1) \text{ Necessary: } (\forall x)(F^*x \supset Mx),$$

which says that it is necessarily true that for any individual x which has F^*—where F^* denotes the total physical state that is responsible for the instantiation of some M-property—the individual also has M. The hidden empirical part in this "law" is of course how the F^*-property itself is obtained: we cannot determine a priori which physical characteristics are "totally" or "maximally" responsible for the instantiation of some mental property M.

Projection clauses like (1) are key ingredients of reductionist proposals and grist on the scientistic mill. They provide the model based on which we are supposed to predict or explain the presence of mental properties, *based on* physical properties of a given entity in a world. We can, as it were, "read off" from a world w's physical properties where and how the mental states and events in w are located and distributed. It follows that interpretive, rational, or phenomenal considerations and explanatory projects, which the conceptual antireductionist claims are necessary for assigning mental properties and identifying mental states and events, can, at least in principle, be circumvented. The existence and distribution of the M-property is fully determined by the F^*-property. Once we have found out which mental properties are necessitated by the F^*-properties, we can predict the mental properties of F^*-occurrences in other worlds. And this strikes defenders of scientism as an attractive model: if we know *how* everything follows from physics—that is, the physical descriptions of worlds—we can see how talk about beliefs, content, and purposes amounts to the production of intriguing narratives, but nothing more than that, because everything is fully determined by physical facts. Or they can, alternatively, argue that if mental descriptions are true, they are true "because they are actually about some non-mental something else, like behavior or brain states" (Antony 2015, 3). But note that this introduces tricky metaphysical questions. If some mental truth is "made true" by physical facts and is, by virtue of those facts alone, reducible to physical truths, scientism seems to depend on a contentious metaphysical conception of truth. Science can suffice with a minimal conception of truth (Horwich 1990) that does not look for an explanation of truth in terms of substantial truth-makers. Scientism shouldn't help itself to contentious metaphysics.

A monist can happily concede that the entity a mental description is true of is going to be numerically identical with something a physical description is true of, just as a psychological description true of me is true of a creature that also falls under physical descriptions. No reductionism follows from this commonsensical fact.

Indiscernibility and Identity

What is the problem with this argument? The problem is that the argument must assume that the physical indiscernibility of two worlds does not collapse into the identity of those worlds. For if two worlds w_i and w_j are identical in the Leibnizian sense when they are physically indiscernible (there are no further nonphysical differences between them), then any talk of projection or determination is the trivial consequence of applying Leibniz's law of the indiscernibility of identicals to worlds. This acknowledges determination, but is consistent with antireductionism. (Reductionism does not follow *trivially* from this argument.) No reductionist argument is required to hold that view, and the burden of the argument shifts to the coherence of conceptual reductionism.

If we allow physics to individuate worlds, as physicalists who defend strong supervenience must hold, then we have indeed provided something stronger than mere physical indiscernibility:[9] if worlds w_i and w_j share all their physical characteristics, then w_i and w_j are identical. Just as two sets are identical when their members are identical, so are two worlds identical when they have the same physical characteristics, that is, when the distribution of physical particles in those worlds is the same. (Physicalism assumes that physics is *closed*: *every* ingredient of w_i falls under a physical description.) But this trivializes the projection claims behind strong supervenience: if an event or object or state e in w_i has a mental property M and w_j has the same physical properties as w_i, then $w_i = w_j$. Hence, whatever further intrinsic or relational mental characteristics e has in w_i, it will have it in w_j. But this can't be the kind of determination that supports scientism, for it is wholly consistent with conceptual antireductionism.

Since strong supervenience is not supposed to be a trivial claim consequence of applying Leibniz's law to physically indiscernible worlds, physical indiscernibility, *as deployed by duplication theorists*, must be prevented from collapsing into identity. At least some physical aspects, namely those determining the physical uniqueness of a world (c.q. the physical uniqueness of an object in a world), must therefore be left out of the projection base. Duplication theorists must exclude from the projection base at least those physical properties that are *uniquely true* of it, or of the world in which it occurs. But then the question arises: why should the physical facts that individuate an event not be required to explain the further mental facts? That would be a very strange assumption, as strange as the idea that our unique location in a city cannot enter into an explanation of what happened to us, and that every explanation which involves our location must appeal to a type of location that can be instantiated by persons who are city-indiscernible.

Here's another way of making the same point. Suppose two superscientists have independently produced what they claim to be a complete physical description of an event in a world—one that individuates that event. Comparing their notes, they discover that they end up with the same description. The best explanation of that happy coincidence is that they were describing not two different worlds but one and the same world. If they insist that the worlds described by their theories are not identical but merely "physically indiscernible," they must either give up their assumption that their descriptions were individuating a world (or an event in a world) *or* accept that nonphysical facts make a difference between both worlds described. But that was explicitly excluded by physicalism.

Duplication theorists and "fixing the further facts" theorists seem to be victims of the same illusion. Talk of physically indiscernible individuals in physically indiscernible worlds suggests that given a substantial supervenience claim and given a physical description of what physical instantiation of a mental characteristic M amounts to, we are free to project M onto physically indiscernible individuals in physically indiscernible worlds. But if physical identity entails numerical identity (that is, if the physical individuates events in worlds), then projection claims are merely trivial consequences of Leibniz's law of the indiscernibility of identicals as applied to worlds, physically individuated. On the other hand, when mental states or events are physically indiscernible yet not identical, the physical characteristics that determine a class of indiscernible entities and enter in projection claims will be less than those that also individuate *uniquely* a particular event. But this indiscernibility claim runs counter to the idea that *every* physical property of an event—including its unique physical properties, the ones that can't be tokened by other events—can be mentally relevant and cannot therefore be excluded from the supervenience basis for mental states (events, objects, processes . . .).

From D-supervenience to Leibnizian Identity

Formally, the principle I introduce is (2), where *F* is a complex physical property that individuates *x* (it specifies the unique physical properties of that event, not a complex physical property that can be shared by two different entities):

$$(2)\ (x)(y)(EF)((Fx \leftrightarrow Fy) \supset (x = y)).$$

Principle (2) says that if any two events *x* and *y* share a uniquely identifying physical characterization *F*, then they must be numerically identical.[10] The contrapositive reading of (2) can be straightforwardly applied to Davidson's own formulation of supervenience: if two events are mentally different, they can't be identical, and so there will be a uniquely identifying physical description true of the first but not of the second. Strong supervenience as required by the brand of physicalism that fosters scientism holds that (2) applies to worlds: if two worlds w_i and w_j are physically indiscernible (if they are individuated by the same physical facts), then those worlds are identical *simpliciter*.[11]

Leibniz's law states that if *x* and *y* are identical, they share all properties:

$$(3)\ (\forall x)(\forall y)(\forall G)((x = y) \supset (Gx \supset Gy)),$$

G being any (physical or nonphysical) property of *x*. From (2) and (3) we can straightforwardly derive that

$$(4)\ (\forall x)(\forall y)(EF)(\forall G)((Fx \leftrightarrow Fy) \supset (Gx \supset Gy)).$$

Davidson's original formulation of supervenience (cf. *supra*) combines Leibniz's law and principle (2). That "there cannot be two events alike in all physical respects but differing in some mental respect" (cf. *supra*) says that if *x* and *y* share physical characteristics that individuate them, they must be identical. It follows, by Leibniz's law, that they share all properties (including the mental properties). By contraposition, if *x* and *y* have different mental properties, then they are (by Leibniz's law) not identical, and it follows from their nonidentity and antecedent monism or physicalism that they *must* have different physical properties in the light of a complete physical theory that is capable of producing their unique physical characteristics. A mental difference thus entails a physical difference.

On this account, we cannot infer a further *systematic* nonphysical difference from the fact that two entities x and y differ in their *individuating* physical characteristics. If x and y are going to differ in their unique physical properties, it may or may not follow that they differ in other characteristics (they may differ biologically, chemically, or mentally). This shows that multiple realizability is *compatible* with [D] and principle (2), but it doesn't capture any deep antireductionist point concerning the mental. Multi-realizability is in a sense too trivial: *any* two physically discernible entities with a common mental characteristic m will have further different physical properties.

On this account, monists who plead for conceptual antireductionism can agree with Rosenberg that physical facts fix all the facts, because they and Rosenberg acknowledge that our best physics will provide individuating descriptions of events, or of worlds in which those events take place. But they differ about conceptual reductionism. Conceptual reductionism, as I have shown, is not required to acknowledge that physical facts fix all the facts, and supervenience of the mental on the physical (a substance) does *not* entail that physics (a set of theories and models) explains everything.

W. V. Quine held that "nothing happens in the world, not the flutter of an eyelid, not the flicker of a thought, without some redistribution of microphysical states." I have explored the contraposition of this claim—that if the total distribution of microphysical states of a world is not redistributed, nothing *else* will change. Ontological monists who defend conceptual antireductionism can be happy with that result, but hard reductionists about the mental must make stronger claims about the explanatory redundancy of our familiar psychological explanations and the idiom they incorporate. An appeal to the supervenience of everything else on the physical is not sufficient to model this point, let alone to provide proof of its truth.

Acknowledgments

I would like to thank the editors of this volume, and especially Maarten Boudry, for discussions and helpful suggestions. This chapter was first presented as a paper, some years ago and under slightly different titles, at conferences in Bielefeld, Germany, and San Sebastian, Spain. I thank Ansgar Beckermann, Jaegwon Kim, and David Sosa for critical comments.

••••••

Notes to Chapter Three

1. Rosenberg 2012 develops this position in greater detail.

2. If every concrete experiential event, process, or state is identical with a physical event, process, or state, and given that identity is a *symmetrical* relation, the label *physicalism* for the doctrine I am about to defend is doubly misleading. It suggests that mental events are "reducible to" physical events (states, processes . . .). But that can't be correct, given identity's symmetry. When it was discovered by Lavoisier that water was H_2O, water wasn't *reduced* to H_2O. When the morning star turned out to be identical with the evening star, the morning star was not *reduced* to the evening star. When heat turns out to be mean molecular motion, heat was not *reduced* to mean molecular motion. Monism *simplifies* our ontology, or at least that part of our ontology relevant for the nature of the domain of the mental.

3. See Van Fraassen 1989 and Davidson 1980 for the interest-relativity of explanations.

4. See Hellman and Thompson 1975.

5. See Davidson 1980 and McDowell 1986. Child (1994) offers an excellent synthesis of Davidson's and McDowell's position. Their arguments will, of course, be rejected by scientistically minded philosophers, but I have already said that this discussion is not the main target of this chapter.

6. The country/city-supervenience example creates Londoners and Parisians, which are still types, not tokens.

7. Accepting physical-to-mental necessities involves the acceptance of such a projection claim.

8. The efforts produced here reveal that the realm of the mental is an essentially open domain, and that therefore no a priori decision can be made on how global supervenience is.

9. See Pietroski 2002, 196 for a similar suggestion and its connections with an actualist metaphysics of "ways the world might be."

10. Compare (2) with Miller's (1990) thesis that every object differs from any other object in some physical respect. See also Hellman and Thompson 1975 for a similar proposal. Hellman and Thompson correctly remark that the principle does *not* entail reducibility of nonphysical predicates to physical predicates and thus leaves room for Davidsonian-inspired antireductionism about psychological predicates.

11. Harry Lewis (1985, 160) points out that "it is a truism that there cannot be two events alike in all physical respects but differing in some mental respect because it is a truism that there cannot be two events alike in all physical respects." Principle (1) formalizes the truism. Note that the assumption works provided that the universe satisfies the principle of the indiscernibility of identicals, although (2) is not to be identified with that principle (it is an individuating principle). See Hacking 1975 for a defense of the principle of identity of indiscernibles, and Pietroski 2002 for further defense that physicalism is a claim about the individuation of worlds in the context of actualism about possibilities.

Literature Cited

Antony, L. 2015. "Defending Folk Psychology." In *Passions and Projections: Themes from the Philosophy of Simon Blackburn,* edited by M. Smith and R. N. Johnson, 3–23. Oxford: Oxford University Press.

Burge, T. 1986. "Individualism and Psychology." *Philosophical Review* 95: 3–46.

Child, W. 1994. *Causality, Interpretation, and the Mind.* Oxford: Oxford University Press.

Davidson, D. 1980. *Essays on Actions and Events.* Oxford: Oxford University Press.

———. 1985. "Reply to Harry Lewis." In Vermazen and Hintikka, *Essays on Davidson,* 242–44.

Grimes, T. 1991. "Supervenience, Determination and Dependency." *Philosophical Studies* 62: 152–60.

Hacking, I. 1975. "The Identity of Indiscernibles." *Journal of Philosophy* 72 (9): 249–56.

Hellman, G., and F. W. Thompson. 1975. "Physicalism: Ontology, Determination, Reduction." *Journal of Philosophy* 72: 551–64.

Horwich, P. 1990. *Truth.* Oxford: Blackwell.

Lewis, D. 1983. "New Work for a Theory of Universals." *Australasian Journal of Philosophy* 61: 343–77.

———. 1986. *On the Plurality of Worlds.* Oxford: Blackwell.

Lewis, H. 1985. "Is the Mental Supervenient on the Physical?" In Vermazen and Hintikka, *Essays on Davidson,* 159–72.

McDowell, J. 1986. "Functionalism and Anomalous Monism." In *Mind, Value and Reality,* pp. 325–40. Cambridge, MA: Harvard University Press.

Miller, R. 1990. "Supervenience Is a Two Way Street." *Journal of Philosophy* 1990: 695–701.

Owens, J. 1989. "Levels of Explanation." *Mind* 98: 59–79.

Papineau, D. 1993. *Philosophical Naturalism.* Oxford: Oxford University Press.

Perry, J. 2002. *Knowledge, Possibility, and Consciousness.* Stanford, CA: CSLI Publications.

Pietroski, P. 2002. *Causing Actions.* Oxford: Oxford University Press.

Putnam, H. 1975. "The Meaning of 'Meaning.'" In *Mind, Language and Reality,* 215–71. Cambridge: Cambridge University Press.

Quine, W. V. 1953. "Identity, Ostension and Hypostasis." In *From a Logical Point of View,* 65–79. New York: Science Library, Harper and Row.

Quine, W.V. 1981. *Theories and Things.* Cambridge, MA: Harvard University Press.

Rosenberg, A. 2011. *The Atheist's Guide to Reality: Enjoying Life without Illusions.* New York: W. W. Norton.

———. 2015. "Disenchanted Naturalism." *Kritikos: An International and Interdisciplinary Journal of Postmodern Cultural Sound, Text and Image* 12 (April), http://intertheory.org/rosenberg.htm.

Strawson, G. 2008. "Real Materialism." In *Real Materialism and Other Essays,* 281–305. Oxford: Clarendon Press.

van Fraassen, B. C. 1989. *Laws and Symmetry.* Oxford: Clarendon Press.

4

Two Cheers for Scientism

TANER EDIS

Pushing Science Too Far

Scientists—particularly those of us who practice a mature science such as physics—tend to be impressed by the achievements of our sciences. Our equipment allows us to probe nature ever more deeply; our theories tie together an increasing amount of knowledge about how things work. Many of us expect not just that science will continue to expand its horizons, but also that common themes and fruitful linkages will appear across many domains of study. Biology will become more mathematical; the social sciences will improve by incorporating ideas from studies of complex systems.

Such ambitions, however, are more controversial when they reach beyond natural science, and even more so when scientists hint that a more scientific approach may help improve the humanities. Many physicists and philosophers are attracted to the notion that everything in our world is made up of the sorts of entities dealt with by physics (Edis 2002; Melnyk 2003). The culture of physics is full of sentiments, usually attributed to prominent physicists, exalting that discipline at the expense of others. Ernest Rutherford has said, "All science is either physics or stamp collecting." But physicists have not stopped there. For example, they have publicly doubted the value of philosophy. Richard Feynman may or may not have said that "philosophy of science is about as useful to scientists as ornithology is to birds," but we do not have to rely just on poorly attributed remarks. In a recent interview, Lawrence Krauss suggested that physics has encroached on philosophical territory:

> Philosophy used to be a field that had content, but then "natural philoso-phy" became physics, and physics has only continued to make inroads. Every time there's a leap in physics, it encroaches on these areas that philos-ophers have carefully sequestered away to themselves, and so then you have this natural resentment on the part of philosophers. . . . Philosophy is a field that, unfortunately, reminds me of that old Woody Allen joke, "those that can't do, teach, and those that can't teach, teach gym." And the worst part of philosophy is the philosophy of science; the only people, as far as I can tell, that read work by philosophers of science are other philosophers of science. It has no impact on physics what so ever, and I doubt that other philoso-phers read it because it's fairly technical. And so it's really hard to under-stand what justifies it. And so I'd say that this tension occurs because people in philosophy feel threatened, and they have every right to feel threatened, because science progresses and philosophy doesn't. (Quoted in Anderson 2012)

These are merely informal comments. They are not, however, rare: they highlight real attitudes within the culture of physics. And they indicate that some physicists, at least, think that science sets the standards for knowledge. They appear to suggest that disciplines such as philosophy, if they do not deal with empirical problems, or if they do not at least help advance the clearly scientific forms of knowledge, also forfeit their claims to producing any sort of genuine knowledge.

Practitioners of disciplines that face criticism from physicists respond de-fensively, often by charging that the physicists' arrogance is symptomatic of scientism (Pigliucci 2015). Perhaps some knowledge is beyond the bounds of science—it is acquired and evaluated in ways that are not continuous with the poking at objects and constructing explanatory models engaged in by natural scientists. Perhaps when some physicists dismiss disciplines outside the sci-ences, they are not merely acting out of ignorance but have a constricted con-ception of knowledge. In some scholarly circles, *reductionism* has become a term of abuse, suggesting a lack of appreciation of complexity, but also hinting at an *illegitimate* attempt at explanation. Similarly, an accusation of scientism often connotes illegitimacy rather than a more ordinary failure of explanation.

Naturally, charges of scientism are part of the defensive rhetoric protecting "other ways of knowing" from science-based criticism (see Law, this volume). After all, Lawrence Krauss took potshots at philosophy while promoting a book criticizing claims that the universe was created supernaturally (2012). Traditional religions have been severely criticized for their commitments to supernatural claims that are historically and scientifically very likely false. But if the fundamental ideas of religion—such as the existence of any gods—are at home strictly within the armchair metaphysical disputes that characterize the traditional philosophy of religion, then physical explanations of the origins of our space-time, such as those favored by a cosmologist such as Krauss, might be irrelevant to the doctrine of divine creation. Many scientists and philosophers are partial to naturalistic views that deny the literal truth of claims of supernatural agency. Some of us go so far as to argue that we live in a world of chance and necessity, that all we know and have any prospect to encounter can be accounted for by physical explanations (Edis and Boudry 2014). But if supernatural agents are revealed by processes internal to religion and discontinuous with science, perhaps such varieties of naturalism are forms of scientism. If supernatural realities are intuited through our response to a holy book, the life of a saint, or our dissatisfaction with the lack of depth of a worldly life, then scientific critiques of supernatural faiths might not be wholly legitimate.

Opposition to suspected scientific overreach can turn into explicit attacks on science. Curtis White, for example, expresses a full-blown romantic rebellion against the claims of modern science:

> The New Atheists speak on behalf of science just as the neuroscientists do, and the message of both camps is: submit. Confess to the superiority of science and reason. But it is not only to evangelicals that this directive is sent; it is also sent to another historical adversary—art, philosophy, and the humanities. There the directive goes something more like this: the human mind and human creations are not the consequence of something called the Will, or inspiration, or communion with a muse or daemon, and least of all are they the result of genius. All that is nebulous; it is the weak-minded religion of the poets. The human mind is a machine of flesh, neurons, and chemicals. With enough money and computing power the jigsaw puzzle

of the brain will be completed, and we will know what we are and how we should act. (White 2013, 8–9)

White argues that most scientists and science popularizers promote an "Enlightenment story" that "is, in its essence, science as ideology (or 'scientism,' as it is often called)":

Unfortunately, scientism takes its too-comfortable place in the broader ideology of social regimentation, economic exploitation, environmental destruction, and industrial militarism that for lack of a better word, we still call capitalism. (White 2013, 10–11)

Science serves political and commercial power, and this is not an accident: "for science the perspectives offered by philosophy, poetry, art, and certainly any kind of spirituality don't exist. For science, the idea that nature, humans, and even formerly intimate things like creativity are all mechanical goes without saying" (White 2013, 109). Moreover, today's scientific enterprises like cognitive neuroscience are thoroughly identified with an overreaching scientism, and are blind to their own incoherence. "The greatest problem with scientism—science's old faith in its jigsaw approach to reality—is that its conclusions about an objective world presuppose a presence—an experiencing thing—that it cannot bring itself to acknowledge. At best, it can try to *persuade* us that this subjective realm of experience is only another kind of object, a chemical machine called the brain whose 'secrets' and 'tricks' we are slowly discovering" (163).

Krauss is wrong in thinking that physicists can dispense with philosophy. But then, critics of science who are very free with accusations of scientism, like White, are also wrong. It is defensible to claim that scientific, philosophical, and humanistic forms of knowledge are continuous, and that a broadly naturalistic description of our world centered on natural science is correct (see Boudry, Blackford, this volume). At the very least, such views are *legitimate*—they may be mistaken, but not because of an elementary error, a confusion of science with an ideology, or an offhand dismissal of the humanities. Those of us who argue for such a view are entitled to have two cheers for an ambitious conception of science; and if that is scientism, so be it. A third cheer would be too much. Not all forms of intellectual life can be assimilated into science, however broadly understood. But that is because we often have other intellectual purposes besides investigation and explanation.

A Loud Cheer: The Case of Psychical Research

Claims of psychic phenomena have long been scientifically investigated. For skeptics, it appears as a case of scientific methods applied to extraordinary claims, which then fail to produce convincing evidence. Tests of psychokinesis or precognition result in tiny effect sizes that further diminish as experimental procedures are made more rigorous. At any given moment, there typically are a few experiments that appear to demonstrate anomalies beyond our understanding of physics. But such experiments invariably turn out to be flawed, inconsistent, or unrepeatable. We find no psychic signal that rises above the noise due to the inevitable complexities and imperfections of our experiments. Spectacular experiences like mystical visions or out-of-body experiences generate no confirmable knowledge and appear to be good candidates for explanation in terms of brain processes in extreme circumstances. The dead don't communicate. Parapsychology has produced, skeptics say, some minor knowledge about nonmagical aspects of psychology, plus a history of experience with how experimental designs are never perfect. Some of the knowledge we have gained about ordinary psychology concerns mistakes, self-deception, and even fraud (Edis 2006).

Parapsychologists have often responded to their rejection by mainstream science by claiming, "Science, which fought for centuries to free itself from the dogma of the Church, is now mired in its own dogma, scientism" (Parra 2013, 15). According to Charles Tart, mainstream science has ossified into scientism. A scientistic commitment to materialism leads skeptics to dismiss all spiritual phenomena that suggest that minds are independent of bodies (1992; 2009). Indeed, skeptics have an overly narrow conception of science. Psychic phenomena are supposed to be linked to minds and spirits: the things that romantics suspect cannot be reduced to material processes. Perhaps skeptical criticism, including tighter experimental protocols, inhibits psychic performances. The sterility of a lab environment, and the boredom inherent in large numbers of repetitions of simple tasks, might work against the artistic, creative subjects who are more likely to be psychically sensitive. The methods of mainstream science are appropriate for investigating material entities. Since parapsychology is an antimaterialist research program, it will challenge the methods of mainstream science as well as its settled conclusions. Skeptics invoke a fixed scientific method, derived from practices successful within the physical sciences. But why should we restrict ourselves to such a method? Assuming that we can apply a predetermined set of procedures to test spiritual phenomena and then pronounce them unreal sounds like a form of scientism.

Skeptical scientists and philosophers are not impressed with such arguments: they appear to be immunizing strategies that protect a failing research program (Boudry and Braeckman 2010, 2012). At face value, psychic powers are supposed to involve extraordinary ways of performing ordinarily observable feats. And these performances have not been forthcoming. Nonetheless, the challenge to scientific methods is not entirely empty. Science does not proceed by a checklist: repeatable experiments, statistical analysis, mathematical predictions, and so on. There cannot be any such thing as a method that produces reliable knowledge in every possible circumstance. What is a good or a bad method for investigation is a fact about the sort of world we live in, like facts about good methods for building bridges. We have a diversity of overlapping methods that change as we develop new capabilities and subject older methods to criticism (See Kitcher, Ross, this volume). And just like bridge building depends on and tests some of our theories of physics, our methods of investigation depend on background theories about how the world works. When we poke at an object of study, we want to be open to being surprised. But we do not even know what we are doing without the help of some theories about the world, including theories about how scientific methods work.

Indeed, as might be expected, the history of psychical research is also a history of debate about and refinement of methods. Researchers recognized the importance of double-blind procedures, tighter precautions against fraud, and some semblance of theory to guide experimental work. And though parapsychological efforts continue, the result so far has been a state of experimental stagnation and a near absence of theoretical explanation (Churchland 1987) that has not changed for decades. Therefore, objections to mainstream scientific methods seem more like evasions than proposals that might productively expand our capabilities for investigation.

More important, the course of events could have gone otherwise. Psychic claims have long included clairvoyance. Parapsychologists have had their desultory studies of "remote viewing," but long before that, visionaries such as Emmanuel Swedenborg produced reports about other planets. The heyday of psychical research included efforts at "occult chemistry," attempting to psychically divine the structure of atoms and compounds (Besant and Leadbeater 1919). All of these failed. But it is not difficult to imagine that occult chemistry could have succeeded. Psychic sensitives could have consistently produced results that were confirmed, were novel, and led to productive research in physical science. The scientific community could then have developed a theoretical understanding. In this alternative reality, an analogue of quantum mechanics really would undergird mystical transfers of information, and physicists rather than new age gurus would talk about the minds of psychics coming into resonance with the structures they visualize. The equipment of physicists would include psychics suspended in sensory deprivation tanks, visualizing the dynamics of quark-gluon plasmas or intuiting the details of a phase transition to a superconducting state. Clairvoyance would be considered a reliable, indispensable method of investigation.

Scientists and philosophers who defend an ambitious, science-centered view of our world—people who get accused of scientism—typically reject psychic powers. If it is scientism to subject psychic claims to reality tests continuous with established forms of scientific knowledge, then scientism is not entirely a bad thing. But it is also clear that such skepticism does not proceed directly from purely scientific practices, if it even makes sense to speak of such things. The facts that we care about include facts concerning good methods of investigation. These facts are not simply embedded in natural science, and they are not trivial. As with any investigation, we start with everyday, unreflective notions of facts, subject them to criticism, and if we are lucky, we end up with a positive feedback between theoretical reflection and reality testing that gives us increased confidence that we have a handle on the world. Traditionally, such reflection on the ways of doing science has been the domain of science studies: the philosophy, history, and sociology of science.

Indeed, it is particularly odd for theoretical physicists such as Lawrence Krauss to denounce the philosophy of science, as they should be most aware of how vital theory is to developing reliable knowledge. At its best, philosophy of science is not the navel-gazing of physicists' caricatures but a disciplined theoretical inquiry into how our sciences work. For reality checks, we confront models of science with examples from the history of science; we critically evaluate our basic ideas about how science is to be done in the process of learning what succeeds (Thagard 1988; Maxwell 2014, 26–31). We ask questions about how our institutions of inquiry can be best structured. As with psychical research, we hope to learn from our successes and mistakes in order to refine our methods. Philosophy, here, is neither window dressing on the obvious progress of science nor an external judge highhandedly laying down the rules defining a scientific method. It is integral to the broader scientific enterprise. In other words, philosophy of science is a vital part of what we might call a science of science. And denying the value of critical reflection on our sciences is not so much scientistic as deeply anti-intellectual.

With a mature science like physics, the role of philosophy is less visible. But younger sciences, like physics itself a few centuries ago, are full of methodological discussions. Practitioners have to regularly stand back and ask metascientific questions about their methods. They aim for normative results, asking what they should do to improve the learning process. In however limited a context, they do philosophy of science. Their efforts may well be improved with closer contact with the broader enterprise of the philosophy of science. Furthermore, even a science such as physics regularly enters more traditionally philosophical territory. For example, the decades-long effort to construct string theory has produced almost no concrete testable results, while remaining mathematically interesting and theoretically compelling to most high-energy theorists. Philosophers and physicists have begun to speak of a postempirical science and "non-empirical theory confirmation" (Dawid 2013). Either to criticize or to defend the scientific status of string theory, physicists have entered debates that look very much like philosophical reflections on the nature of scientific theories (Smolin 2006; Ellis and Silk 2014).

There is nothing illegitimate about broad-ranging ambitions for science. But pushing science as far as it can go depends on recognizing the continuity between science and philosophy (Boudry 2013; Haack 2007). There is no sharp separation between philosophical questions and theoretical questions in science. Science, broadly understood, is not fixed. It includes criticism of methods, and so it is capable of adapting to new circumstances. And therefore, in cases such as psychical research, an accusation of scientific overreach does not have much traction. If we want to figure out whether psychic phenomena are real, broadly scientific criticism is exactly the right tool for the job.

A Mild Cheer: The Case of Alternative Medicine

At first glance, the case of alternative medicine may look much like that of psychic phenomena. There is, in fact, considerable overlap: forms of alternative medicine include psychic and spiritual healing, as well as homeopathy, acupuncture, reflexology, and so forth. And as with psychic phenomena, claims of alternative medicine often contradict well-established physics, and tightly controlled research into the more magical alternative healing modalities has typically produced ambiguous or negative results. Proponents of evidence-based medicine (EBM) argue that almost all alternative healing methods fail to produce adequate scientific evidence for efficacy (Ernst et al. 2001). Homeopaths or spiritual healers cure neither cancer nor the common cold.

Also similarly to the case of psychical research, defenders of alternative medicine within the medical professions criticize the methods of testing alternative claims, arguing, for example, that EBM's demand for randomized controlled trials arbitrarily imposes a philosophy alien to the ways of knowing claimed by alternative healing modalities (Tonelli and Callahan 2001). Accusations of scientism appear often. For example, Andrew Miles (2009) blames the "scientistic reductionism" exemplified by EBM for problems in modern medicine such as the lack of an ethic of care and a devaluing of clinical judgment: "The advent and rise of EBM codifies the modern scientism in medicine, then, and demonstrates all of the characteristics of scientism: radical reductionism, the privileging of the scientific method and inquiry above all others and a marked tendency to totalitarianism, even microfascism." Alternative practitioners, such as the homeopath Lionel Milgrom, warn that "in the United Kingdom, scientism is on the march, attempting to crush beneath its positivist boot the public's right to complementary and alternative medical therapies such as homeopathy" (2011, 34). He denounces

an insidious "crypto-fascist" brew of scientism, globalised corporatism, and (ironically, originally left-wing) militant materialistic nihilism which denies humanity has any inner metaphysical reality. Thus, it reduces human beings to biochemical mechanisms, from which it is but a short step to regarding humans as nothing other than biologically-determined economic units. This suits globalised conglomerates very well which, via their encroaching control over individual governments, would no doubt wish to see representative democracy reduced to the irrelevance of some inane reality game show. (Milgrom 2011, 36)

Now, all this, including the overheated talk about fascism, again looks very much like elements of an immunizing strategy, protecting alternative medicine from its critics. The complication, however, is that the appeal of alternative medicine is not just due to straightforward fact claims about curing or reducing the severity of organic diseases. As Kaptchuk and Eisenberg observe:

The attraction of alternative medicine is related to the power of its underlying shared beliefs and cultural assumptions. The fundamental premises are an advocacy of nature, vitalism, "science," and spirituality. These themes offer patients a participatory experience of empowerment, authenticity, and enlarged self-identity when illness threatens their sense of intactness and connection to the world. (Kaptchuk and Eisenberg 1998, 1061)

Alternative medicine often appeals to sufferers because of the quasi-religious context in which it locates disease. Alternative healers promise to situate the "whole person" in a cosmic drama, as opposed to industrialized biomedical practices that treat patients as bearers of diseased systems to be fixed through the biological analogue of automotive repair. They help alter the patient's perception of self and the meaning of her suffering. Indeed, historically the concept of healing has usually been more expansive than what can be achieved through medical interventions. It has often included integration with a community and has been connected to an overall vision of human purpose in life. Therefore, the practice of medicine has been, until recently, more an art than an applied science, with overtones of priesthood rather than biotechnological wizardry.

Even today, the aims of medical care remain broader than the investigation and explanation that are the focus of biomedical science. And medicine is not just an applied science: for medicine, science is an important tool in the service of complex social and individual purposes. For many of their enthusiasts, the success of alternative modalities of healing go beyond a placebo effect or the various ways we all find to delude ourselves into thinking the money we spent was not wasted. The more functionally religious forms of alternative healing ideologies redefine, in part, what the aims of the patient are, what "success" means. Even when their cancer persists, patients may reconcile themselves to the disease by finding a new meaning for this calamity in their lives.

This is not to say that alternative healing ideologies are immune to criticism, since in most cases, while their claims to physically cure disease may become attenuated, such claims do not disappear. The patients' hopes are not entirely met. And many of the shortcomings of medical practice exploited by alternative medicine, such as the impersonal, uncaring, and financially exploitative aspects of many patients' encounters with modern health care systems, can conceivably be addressed without sacrificing EBM. Therefore, accusations of scientism still, in this case, largely function to protect factual claims from criticism.

Nonetheless, medicine has aims other than just the production of relevant knowledge. Usually, these aims reinforce each other: a better understanding of cancer mechanisms will help the relief of suffering, or reduce economic losses due to disease. But with finite resources, such different aims can also compete with each other; if EBM is understood to be a translation of the best biomedical evidence into clinical practice, this raises broader social and political questions (Mykhalovskiya and Weir 2004). Inevitably, biomedical science, with its primary interest in investigation and explanation, must slightly diverge from the larger social enterprise of health care. If scientism implies a kind of institutional imperialism, with the aims and practices of natural science taking over other institutions, it cannot succeed completely even in the case of an applied science such as medicine.

We may still muster a cheer for scientism when we consider the question of medical knowledge. EBM, or science-based medicine more broadly (Gorski and Novella 2014), still appears to do the best job of investigating and explaining disease processes. It is not perfect, but no "other ways of knowing" are intellectually serious rivals. But while scientism is often understood to be a claim to assimilate all knowledge into natural science, it may be worth paying more attention to the normative aspects of alleged examples of scientism. EBM is not just a desire to attend to the best evidence while preparing medical textbooks. It also suggests that science should determine clinical practice, and even public policy. Skeptics of psychic and supernatural claims criticize these claims not just because they look false but because they often think that widespread acceptance of such claims has more important adverse consequences than just a failure to legitimately satisfy curiosity. They usually *assume* a secular, liberal political outlook: as autonomous individuals, we decide on our personal purposes in life, but we also should use the best of our knowledge to pursue our collective aims. Hence creationism, astrology, psychic healing, or homeopathy must be recognized as mistakes in the public realm. Disputes about scientism are usually also arguments about the *political* role of science. It is no accident that defenders of ways of knowing that are rejected by natural science, or are supposed to be discontinuous with a broader scientific and philosophical enterprise, so readily associate scientism with political repression.

As philosophically interesting as a scientistic approach to knowledge may be, therefore, the political aspects of debates over scientism need more emphasis. The knowledge claimed when we say that someone is knowledgeable about cheese may only be distantly continuous with science, much less formalized, much more mixed up with purposes other than explanation and investigation. But people accused of scientism rarely object to such everyday forms of expertise. The knowledge claimed by cheese connoisseurs does not violate a naturalistic, science-based picture of our world. There are no powerful institutions that impose a correct way to appreciate cheese.

Health care, particularly public health, is another matter. In a secular, liberal political setting, it matters that we get public knowledge about health right. Our political tradition defers to science as an exemplary form of public reason in exactly such circumstances. We have much better reasons to trust that science-based medicine is accurate than we have for any of its alternatives. This political context makes the case of alternative medicine murkier than the case of psychic phenomena, which was limited to questions of fact in a narrow, scientific institutional setting, with far less negotiation of competing interests. There is more to medical care than biomedical science. Nonetheless, scientism is still on the right track in this case.

A Silent Cheer: The Case of Morality

A plurality of aims and a political context complicate the role of science in medicine. But scientism may yet work, if there are moral facts about the correct way to balance different aims, and if moral knowledge can be assimilated into science. Science, broadly conceived, might include moral philosophy, such that a process of investigation and explanation can give us not just what is but also what ought to be.

Among prominent antireligious polemicists, some have expressed exactly such a view: that there are natural moral facts that are objectively true, available to a well-informed rational person, that are binding, universal, and motivating, and that can be discovered through science (Harris 2010, Carrier 2011; for criticism of Harris, see Kalef, this volume). They usually take a naturalistic Aristotelian approach, arguing that morality is about human flourishing. For such authors, then, finding the facts about human flourishing is a matter for an applied science akin to medicine.

Few find such an approach promising. Even for those of us with an ambitious conception of science and a naturalistic understanding of the world, there are serious difficulties with moral facts akin to the facts of chemistry.

Our sciences certainly have much to say about the nature of morality: our prosocial emotions, moral perceptions, options for collective behavior, even how we build systems of morality and engage in moral discourse. Understanding what we do when we practice morality may well call for a broadly scientific enterprise, incorporating metaethical theories in a way similar to how philosophy of science is indispensable to understanding science.

If we approach morality in this manner, we can find, for example, that actors face objectively good or bad strategies to negotiate their social landscapes in order to achieve their aims (Churchland 1996). Our aims, however, are diverse. Instead of a fixed human nature that sets overarching natural aims, we are far more likely to end up with a picture of a "moral ecology" (Flanagan 2002), in which multiple stable, successfully reproducing ways of life occupy niches in a changing social environment. Not just anything goes, just like only a small subset of gene combinations can make a viable organism, and there are only a finite number of evolutionarily stable strategies in play. Nonetheless, different ways of life within a moral ecology will sustain different moral intuitions and behaviors. Human nature is too fluid, and the social landscape too varied, to sustain the sort of universal, binding moral facts required to conceive of morality as a kind of applied science.

If this is so, naturalists with an ambitious view of science will be drawn toward metaethical views resembling amoralism (Garner 1994), or a relativism emphasizing the inseparability of moral demands from actors' various interests and agreements (Harman 2000). Broad commonalities in human nature may moderate such views, but we will still be left with value pluralism. We all value certain things such as freedom, security, and so forth. But such values are plural and incommensurable. Human values inevitably harbor conflict and competition, both among themselves and within themselves (Crowder 2002; Graham et al. 2012). In all such views of morality, science only has a limited role in advancing moral discourse. Our sciences may provide a fuller description of a social landscape, but they cannot reveal moral facts about how to negotiate between competing values or adjudicate between rival ways of life. Moreover, modern science has helped to cast doubt on supernatural anchors for moral convictions. That only emphasizes the problem: with a science-centered view, it appears more likely that there are no hard, universal moral facts to be found.

Even stronger appeals to human nature cannot turn moral questions into problems to be solved by a scientific style of investigation. This is an era of biotechnology, where it has become easier to imagine that we can change human nature; bioethicists have to worry about how and whether we should (Habermas 2003). We can seek answers by referring to our stable interests, or our overarching aims rooted in human nature. But this becomes increasingly difficult as more and more of our interests and our nature become a matter of choice. After all, we would also have the option of changing our personalities to respond to any situation. If we were to acquire low-cost means of sculpting our selves—our personalities, our interests, our nature—it would become impossible to rely on a stable background of interests against which to judge minor alterations of selves. The result is a kind of moral vertigo (Owens 2007). Appealing to human nature does not help with questions about whether we ought to make significant changes in human nature.

If something like this picture of our moral lives is correct, scientism about morality starts to look dubious. On one hand, morality is not associated with an "other way of knowing" separate from scientific knowledge. Morality is an important part of our lives as social animals trying to make our way in the world, but it is inescapably plural, bound up with particular interests and communities. Religious thinkers, even when they accept value pluralism, tend to see it as a problem to be solved by appealing to a supernatural form of moral knowledge (Englehardt 2004). Naturalists think that such an appeal does not help to explain our moral lives. On the other hand, while our sciences aspire to a "view from nowhere" (Nagel 1986), moral judgments are always made within perspectives that include particular interests and aims. Making our way in the world demands more than knowledge; our abilities to investigate and explain are only part of the story.

If we emphasize the political aspect of scientism, focusing on the public, institutional role of science, then scientism looks even odder. Many of us with a high regard for science have a broadly secular liberal political outlook; even liberal philosophers who want to allow religious communities a stronger voice in public life usually insist on deferring to science when facts relevant to policy are at stake (Habermas 2008, 137). Secular liberals balance and shape their various aims in a manner that privileges science when public matters of fact are to be decided. But for other ways of life in our moral ecology, deference to science will not come as easily.

Conservative religion is an obvious example. Far from deferring to scientific institutions, the more fundamentalist forms of religion nurture some of the starkest forms of science denial, such as creationism. Naturally, strategies to immunize creationism from criticism include accusing proponents of evolution of scientism (Morris 2002). Moreover, as with some alternative medical ideologies but even more strongly, conservative religion thrives on shaping the interests and personalities of its adherents. Secular liberals care for personal autonomy, while religious ideals more often prescribe a disciplining of the self, favoring religious virtues such as submission and obedience to sacred authority (Mahmood 2001). A patient adopting an alternative medical practice makes a mistake, to the extent that she retains a strong interest in reducing the effects of her disease. But a religious person training herself to better conform to a list of religious virtues changes her interests. Participating in her community shapes her very personality, so that her aims in life are better aligned with religious ideals. In that case, if creationism helps her to deflect science-based challenges to her ability to pursue her religious aims, then it is much harder to call her creationism a mistake.

Now, almost everyone, including creationists, values cognitive accuracy to *some* extent. Very often this has an instrumental rationale: accuracy helps everyone choose better ways to pursue their interests. This advantage, however, has to be balanced against the *costs* of accuracy. Properly appreciating complex, counterintuitive scientific ideas such as evolution or modern physics demands considerable resources, and so cognitive accuracy competes with other aims in life. The best solution will often be to adopt low-cost, quick-and-dirty cognitive strategies that do well enough most of the time, even if they produce large errors in unusual circumstances removed from everyday concerns. Cognitive scientists have proposed that normal human psychology includes a "hyperactive agency detection device" (Barrett 2000; Lisdorf 2007). Since the costs of failing to detect agents are much higher than the costs of detecting agency when none is present, our quick-and-dirty cognitive mechanisms bias us toward perceiving purpose and agency everywhere, contributing to our susceptibility to belief in supernatural agents. There are also cultural means of reducing the costs of cognition at the expense of some accuracy. For example, deference to established authorities helps share cognitive costs throughout communities.

For many of us, then, cognitive accuracy comes at a high cost, and can be sacrificed in favor of low-cost alternatives that are good enough for everyday purposes. Very few religious communities cripple the ability of their adherents to function in everyday life. Even for those communities that make strong demands that entail withdrawing from modern life into ultraorthodox enclaves, such costly displays of commitment are compensated by access to very strong networks of mutual aid. Religious training and shaping of selves further adjust the costs associated with better accuracy. Certain kinds of supernatural beliefs, even if false, make sense in the context of a religious life, and are often useful in promoting a cohesive community with a strong sense of moral purpose (Atran and Henrich 2010).

Creationists and tight-knit enclaves are perhaps extreme examples, but almost all of us are in similar situations. Almost all of us want to be accurate; even if we adopt a belief such as creationism, it is to protect what we think are important religious truths. However, even if we do not acknowledge it, cognitive accuracy is rarely an overriding aim in our lives; it is more often a means to other ends. And accuracy is always costly, always potentially in conflict with other values. Even our intellectual values are plural. We adopt strategies that make sense in the context of our overall purposes, and some of us, given our positions in our moral ecology, will reasonably value science less than others.

In a complex society, a small minority may still adopt a kind of political scientism, exaggerating secular liberal tendencies to defer to scientific institutions. They may even adopt the useful falsehood that the moral facts as revealed by science demand that science enjoy a kind of imperial intellectual authority. Such a political position, however, stands against the plurality of moral outlooks in complex societies. A secular liberal political framework more typically acknowledges plurality and seeks ways to live together without imposing a comprehensive moral vision on a diverse society (Davion and Wolf 2000). This requires compromises: constraints on strongly religious institutions, but also a limited public role for science. Our political negotiations may result in evolution, not creation, in public school classrooms. But our schools will not teach that Darwinian evolution gives strong reasons to doubt the reality of the gods. The secular liberal tradition emphasizes public reason to help make collective political decisions. But public reason is always contested and circumscribed by the worldly considerations most citizens can agree are up for debate and negotiation. Political scientism, if given full rein, would resemble "scientific socialism" or there-is-no-alternative neoliberalism more than a politics of compromise and pluralism.

Adopting antiscientistic rhetoric that associates political scientism with totalitarianism would go too far. And yet, the totalizing impulse within scientism is real enough. Those of us who have an ambitious conception of science, however, can contain this impulse. We can recognize that our degree of deference to science makes sense within *our* way of life, and that even then, we always have other interests to balance against those that led us to look toward science. We will have disagreements and even conflicts with others, such as the conservatively religious. But we will also have opportunities for compromise and cooperation.

Harmless and Harmful Scientisms

The question of psychic powers can be treated in a straightforwardly scientific context, in which case granting mainstream scientific institutions cognitive authority is not too difficult. With alternative medicine, things get murkier. Political as well as scientific questions become relevant, and the authority we grant the institutions of science becomes more circumscribed. And when debating the political and moral questions in complex, pluralistic societies, the influence of scientific institutions may be even more limited.

In that case, some distinctions may help us assess accusations of scientism. The variety of scientism that has the best claim to respectability concerns continuities in our forms of knowledge. It depends on a very broad understanding of science, including not just the natural and social sciences but those aspects of mathematics, philosophy, and the humanities most concerned with investigation and explanation. This variety of scientism highlights continuities in the various ways we produce knowledge, and weaves the products of our knowledge-seeking enterprises into a naturalistic overall picture. Scientism in this sense will often emphasize natural science as a particularly noteworthy example of success. But by and large, this scientism is harmless: it seeks connections and coherence, not intellectual conquest.

Certainly, even a relatively harmless scientism will be contested. Some will, for example, argue that naturalism, even setting aside its reductionist and scientistic tendencies, ignores spiritual realities and leads to a flattening of our lived experience, even without dire political consequences (Taylor 2007). But in considering such arguments, it does not help to use terms like *scientism* or *reductionism* in a manner that implies illegitimacy. Whatever else it may be, naturalism and its accompanying harmless scientism are intellectually legitimate points of view today.

Now, just as reductionism can lapse into "greedy reductionism" (Dennett 1996), scientism can take on harmful forms. Physicists disrespecting philosophy is a good example. We could call this "philistine scientism": dismissal of what superficially looks woolly and nonscientific without a proper appreciation of how a discipline might produce genuine knowledge. This is no less intellectually lazy than trying to delegitimate criticism by using scientism as an insult.

That being said, the coolness toward philosophy expressed by many physicists is not entirely without reason. Physicists understandably form their opinions of philosophy based on the more publicly available and attention-grabbing philosophical ideas. The last few generations of physicists have now encountered the philosophy of science in the shape of Feyerabendian anarchism and Kuhnian paradigm-talk, followed by the decades of the Science Wars (Parsons 2003; Koertge 2007), and the lingering tendencies toward postmodern posturing and cognitive relativism in the humanities. Scientists who venture to criticize paranormal and supernatural claims run into the philosophy of religion, and discover that it remains a stereotypically armchair-based metaphysical enterprise dedicated to dissecting endless variations on the classical arguments for and against a theistic God. In this state of permanent stalemate, whatever conceptual advances that may be attained are very weak tea compared with the robust sense of progress within the sciences. It has been tempting for many scientists to extend such observations to the conclusion that there is something wrong with philosophy in general.

The remedy to a philistine scientism is to better educate the scientists. Too often, debates over scientism resemble exercises in academic turf protection. Instead, we need more awareness of examples such as the productive interactions between cognitive neuroscientists and philosophers of mind, where researchers often ignore disciplinary boundaries, and there is no clear separation between philosophically interesting and scientifically vital questions. There are plenty of other instances: physicists, for example, could be reminded of the interesting work that is done in the interstices of philosophy and physics concerning the nature of time (Callender 2011). Better working relationships between scientists and philosophers, as encouraged by harmless scientism, is a more promising way to prevent prominent physicists from making unfortunate public statements about philosophy.

Political varieties of scientism are more difficult to describe as harmless. The notion of ordering our lives according to science periodically surfaces in antireligious circles. When coupled with the utopian tendencies within secular politics, political scientism can be an illiberal influence. Our history with attempts at salvation through science should lead to some caution.

Where good science is available, however, many of us with secular liberal political inclinations will want science to inform our public debates. Some of us will hope for a more expansive role for applied scientific institutions, the way we prefer a biomedical approach to health care rather than public health services run by homeopaths. There is a balance to be struck, and this will come about through political negotiation. Indeed, what we need is not just scientific knowledge but a measure of political wisdom, which has usually been better cultivated within the humanities.

In that case, it is also worth asking how much of a problem scientism is. Denunciations of scientism regularly appear in apologetics for dubious claims on the fringes of science or for the social privileges extended to religion. In response, scientists who criticize supernatural claims and who are accused of scientism often conclude that the term *scientism* is too vague, and of little use but as evasive rhetoric (Coyne 2015). In such contexts, scientism appears to be harmless: we can pronounce two cheers for scientism and get on with the work of criticism.

In other situations, however, a more political scientism is not quite so innocent. Those of us who are secular liberals often have good reason to wish for a deeper appreciation of science; after all, we face catastrophic climate change that has been aided and abetted by undue distrust of natural science. But deference to science also has its dangers: sometimes, the claim of a scientific status, with the implication of superiority to other intellectual enterprises, can serve to protect dubious ideas from criticism. There might, for example, be a case that the neoliberal political orthodoxies of today are bolstered by presenting neoclassical economics as a science, and that this reputation is not well earned (Keen 2011).

More fundamentally, secular liberals especially cannot afford to devalue intellectual enterprises that are not easily assimilated into a framework of applied science. Our political ideals are closely linked to our aspirations for liberal education. But today, public education, especially in the humanities, faces austerity and privatization (Newfield 2008). One reason is the perception of the less "hard," less "scientific" forms of scholarship as being less useful in terms of economic return or production of human capital. But such a political scientism harbors a threat to basic science as well as the humanities. After all, many an astrophysicist or evolutionary biologist will find it no less difficult to justify their work in such narrowly pragmatic terms.

If any variety of scientism is to be defensible, it must, above all, be harmless. If other forms of inquiry are continuous with science, this also means that natural science is, in part, a way of doing philosophy, and that scientific writing is a form of literature. For those of us who practice and admire natural science, this should not be a problem. Science is not diminished when we recognize that it is but one part of a much larger intellectual enterprise.

······

Literature Cited

Anderson, Ross. 2012. "Has Physics Made Philosophy and Religion Obsolete?" *Atlantic*, April 23, http://www.theatlantic.com/technology/archive/2012/04/has-physics -made-philosophy-and-religion-obsolete/256203/.

Atran, Scott, and Joseph Henrich. 2010. "The Evolution of Religion: How Cognitive By-Products, Adaptive Learning Heuristics, Ritual Displays, and Group Competition Generate Deep Commitments to Prosocial Religions." *Biological Theory* 5 (1): 18–30.

Barrett, Justin L. 2000. "Exploring the Natural Foundations of Religion." *Trends in Cognitive Sciences* 4 (1): 29–34.

Besant, Annie, and Charles W. Leadbeater. 1919. *Occult Chemistry: Clairvoyant Observations on the Chemical Elements.* London: Theosophical Publishing House.

Boudry, Maarten. 2013. "Loki's Wager and Laudan's Error: On Genuine and Territorial Demarcation." In *Philosophy of Pseudoscience: Reconsidering the Demarcation Problem*, edited by Massimo Pigliucci and Maarten Boudry, 79–98. Chicago: University of Chicago Press.

Boudry, Maarten, and Johan Braeckman. 2010. "Immunizing Strategies and Epistemic Defense Mechanisms." *Philosophia* 39: 145–61.

———. 2012. "How Convenient! The Epistemic Rationale of Self-Validating Belief Systems." *Philosophical Psychology* 25 (3): 341–64.

Callender, Craig, ed. 2011. *The Oxford Handbook of Philosophy of Time.* Oxford: Oxford University Press.

Carrier, Richard. 2011. "Moral Facts Naturally Exist (and Science Could Find Them)." In *The End of Christianity*, edited by John W. Loftus, 333–58. Amherst, NY: Prometheus Books.

Churchland, Paul M. 1987. "How Parapsychology Could Become a Science." *Inquiry* 30 (3): 227–39.

———. 1996. "The Neural Representation of the Social World." In *Mind and Morals: Essays on Cognitive Science and Ethics*, edited by Larry May, Marilyn Friedman, and Andy Clark, 91–108. Cambridge, MA: MIT Press.

Coyne, Jerry A. 2015. *Faith versus Fact: Why Science and Religion Are Incompatible.* New York: Viking.

Crowder, George. 2002. *Liberalism and Value Pluralism.* London: Continuum.

Davion, Victoria, and Clark Wolf. 2000. *The Idea of a Political Liberalism: Essays on Rawls.* Lanham, MD: Rowman and Littlefield.

Dawid, Richard. 2013. *String Theory and the Scientific Method.* Cambridge: Cambridge University Press.

Dennett, Daniel C. 1996. *Darwin's Dangerous Idea: Evolution and the Meanings of Life.* New York: Simon and Schuster.

Edis, Taner. 2002. *The Ghost in the Universe: God in Light of Modern Science.* Amherst, NY: Prometheus Books.

———. 2006. *Science and Nonbelief.* Westport, CT: Greenwood Press.

Edis, Taner, and Maarten Boudry. 2014. "Beyond Physics? On the Prospects of Finding a Meaningful Oracle." *Foundations of Science* 19 (4): 403–22.

Ellis, George, and Joe Silk. 2014. "Defend the Integrity of Physics." *Nature* 516: 321–23.

Englehardt, H. Tristram, Jr. 2004. "Taking Moral Difference Seriously: Morality after the Death of God." In *Recognizing Religion in a Secular Society: Essays in Pluralism, Religion, and Public Policy*, edited by Douglas Farrow, 116–39. Montreal: McGill-Queen's University Press.

Ernst, Edzard, Max H. Pittler, Clare Stevinson, and Adrian White, eds. 2001. *The Desktop Guide to Complementary and Alternative Medicine: An Evidence-Based Approach.* London: Harcourt.

Flanagan, Owen. 2002. *The Problem of the Soul: Two Visions of Mind and How to Reconcile Them.* New York: Basic Books.

Garner, Richard. 1994. *Beyond Morality.* Philadelphia: Temple University Press.

Gorski, David H., and Steven P. Novella. 2014. "Clinical Trials of Integrative Medicine: Testing Whether Magic Works?" *Trends in Molecular Medicine* 20 (9): 473–76.

Graham, Jesse, et al. 2012. "Moral Foundations Theory: The Pragmatic Validity of Moral Pluralism." *Advances in Experimental Social Psychology* 47: 55–130.

Habermas, Jürgen. 2003. *The Future of Human Nature.* Cambridge: Polity Press.

———. 2008. *Between Naturalism and Religion: Philosophical Essays.* Translated by Ciaran Cronin. Malden, MA: Polity Press.

Haack, Susan. 2007. *Defending Science—within Reason: Between Scientism and Cynicism.* Amherst, NY: Prometheus Books.

Harman, Gilbert. 2000. *Explaining Value and Other Essays in Moral Philosophy.* New York: Oxford University Press.

Harris, Sam. 2010. *The Moral Landscape: How Science Can Determine Human Values.* New York: Free Press.

Keen, Steve. 2011. *Debunking Economics—revised and Expanded Edition: The Naked Emperor Dethroned?* London: Zed Books.

Koertge, Noretta. 2007. "The Science Wars." *Minerva* 45 (1): 105–11.

Krauss, Lawrence M. 2012. *A Universe from Nothing: Why There Is Something Rather Than Nothing.* New York: Free Press.

Kaptchuk, Ted J., and David M. Eisenberg. 1998. "The Persuasive Appeal of Alternative Medicine." *Annals of Internal Medicine* 129 (12): 1061–65.

Lisdorf, Anders. 2007. "What's HIDD'n in the HADD?" *Journal of Cognition and Culture* 7 (3): 341–53.

Mahmood, Saba. 2001. "Feminist Theory, Embodiment, and the Docile Agent: Some Reflections on the Egyptian Islamic Revival." *Cultural Anthropology* 16 (2): 202–36.

Maxwell, Nicholas. 2014. *Global Philosophy: What Philosophy Ought to Be.* Exeter, UK: Imprint Academic.

Melnyk, Andrew. 2003. *A Physicalist Manifesto: Thoroughly Modern Materialism.* New York: Cambridge University Press.

Miles, Andrew. 2009. "On a *Medicine of the Whole Person*: Away from Scientistic Reductionism and Towards the Embrace of the Complex in Clinical Practice." *Journal of Evaluation in Clinical Practice* 15: 941–49.

Milgrom, Lionel. 2011. "Beware Scientism's Onward March!" *American Journal of Homeo-pathic Medicine* 104 (1): 33–39.

Morris, Henry M. 2002. "What Are They Afraid Of?" *Acts and Facts* 31 (12): a–c.

Mykhalovskiya, Eric, and Lorna Weir. 2004. "The Problem of Evidence-Based Medicine: Directions for Social Science." *Social Science and Medicine* 59: 1059–69.

Nagel, Thomas. 1986. *The View from Nowhere*. New York: Oxford University Press.

Newfield, Christopher. 2008. *Unmaking the Public University: The Forty-Year Assault on the Middle Class*. Cambridge, MA: Harvard University Press.

Owens, David. 2007. "Disenchantment." In *Philosophers without Gods: Meditations on Atheism and the Secular Life*, edited by Louise Anthony, 165–78. New York: Oxford University Press.

Parra, Alejandro. 2013. "2012 Presidential Address: What Have We Learned about Psi? Reflections on the Present of Parapsychology." *Journal of Parapsychology* 77: 9–20.

Parsons, Keith M., ed. 2003. *The Science Wars: Debating Scientific Knowledge and Technology*. Amherst, NY: Prometheus Books.

Pigliucci, Massimo, ed. 2015. *Scientific Chronicles: Exploring the Limits, If Any, of the Scientific Enterprise*. New York: ScientiaSalon.org.

Smolin, Lee. 2006. *The Trouble with Physics: The Rise of String Theory, the Fall of a Science, and What Comes Next*. Boston: Houghton Mifflin.

Tart, Charles T. 1992. "Perspectives on Scientism, Religion, and Philosophy Provided by Parapsychology." *Journal of Humanistic Psychology* 32 (2): 70–100.

———. 2009. *The End of Materialism: How Evidence of the Paranormal Is Bringing Science and Spirit Together*. Oakland, CA: New Harbinger Publications.

Taylor, Charles. 2007. *A Secular Age*. Cambridge, MA: Belknap Press of Harvard University Press.

Thagard, Paul. 1988. *Computational Philosophy of Science*. Cambridge, MA: MIT Press.

Tonelli, Mark R., and Timothy C. Callahan. 2001. "Why Alternative Medicine Cannot Be Evidence-Based." *Academic Medicine* 76 (12): 1213–20.

White, Curtis. 2013. *The Science Delusion: Asking the Big Questions in a Culture of Easy Answers*. Brooklyn, NY: Melville House.

5

. . . .

Scientism and the Is/Ought Gap

JUSTIN KALEF

Scientism, as I will be using the term in this chapter, is the view that the only facts are those that could in principle be learned exclusively from the natural and social sciences, empirical observations, mathematics, and logic; and that any beliefs that can be justified only in some other way are merely sham knowledge. And yet, adherents of scientism often speak and act as though we *ought* to do things (for instance, that we ought to accept scientism). Clearly, if it is a fallacy to derive an ought from an is, then these devotees of scientism are in trouble.

The putative fallacy of deriving an ought from an is, sometimes called the *naturalistic fallacy,* arises from the simple observation that an argument whose conclusion nontrivially contains a concept that does not appear in any of its explicit or implicit premises cannot be valid. An apparently straightforward application of this general principle is that an argument whose conclusion nontrivially contains the concept of ought (or morally right, morally good, etc.), but whose premises do not contain that concept, cannot be valid.

Is it possible that the naturalistic fallacy is *not* in fact a fallacy? While the reasoning behind the fallacy diagnosis seems airtight, there is one possibility that could, in theory at least, make room for a scientistic approach: the possibility that the move from an is to an ought may be legitimate in some cases. As

Charles Pigden (2013) remarks in "The Is-Ought Gap," certain deontic logics may employ logical "ought" operators that appear in the premises without being mentioned in the conceptual content of the statements, but that become more salient in the conclusion. Although it is a matter of considerable controversy which modal logical system, if any, is to be preferred in such cases, and which system of deontic logic, if any, is the correct correlate of that general modal system, it remains a live possibility that an ought can legitimately be derived from an is.[1] I explore this interesting possibility a little later.

Much of the discussion of scientism takes place in popular books and lectures, far away from technical controversies over operators and metatheorems in deontic logics. In these popular discussions, it is common for moral philosophers to be portrayed in caricature as benighted longhairs stumbling around in the dark for want of empirical enlightenment—enlightenment that scientism alone can provide. Sam Harris (2010), perhaps the most notorious among the "scient-ists" in the popular press, has written a book and given several talks in support of the supposedly revolutionary view that science can inform human values and steer us away from moral relativism. The fact that prominent philosophers have been articulating clear moral views, and opposing relativism, for millennia seems not to impress Harris: in his self-presentation, the attempt to use observations—and, curiously enough, *nonempirical* thought experiments such as one involving a fictional society that systematically blinds every third child in obedience to some religious scriptures—represents a new and hitherto neglected direction in moral thinking.

It is hard not to wonder why Harris thinks his approach of considering the overall levels of happiness in objective or hypothetical cases is some sort of novelty, or why he would brazenly present this view with great fanfare as a significant achievement if he recognized that it is what most philosophers, and perhaps most people, already employ and have perhaps always employed in working through moral issues. It is also tempting to wonder whether Harris really grasps the nature of the is/ought problem, as when he says:

> [G. E.] Moore felt that his "open question argument" was decisive here: it would seem, for instance, that we can always coherently ask of any state of happiness, "Is this form of happiness itself good?" The fact that the question still makes sense suggests that happiness and goodness cannot be the same. I would argue, however, that what we are really asking in such a case is "Is this form of happiness conducive to (or obstructive of) some higher happiness?"

This question is also coherent, and keeps our notion of what is good linked to the experience of sentient beings. (Harris 2005, 283–84)

Unfortunately, Harris does not tell us what his argument for that conclusion would be in the unspecified circumstances under which he "would argue" for it. But to be as fair as possible to Harris, I focus on what I take to be his clearest articulation of how he thinks it possible to cross the is/ought divide. His plan is to make the crossing via an articulation of the synonymy of certain moral and nonmoral terms:

> To say we "should" follow some of these paths and avoid others is just a way of saying that some lead to happiness and others to misery. "You shouldn't lie" (prescriptive) is synonymous with "Lying needlessly complicates people's lives, destroys reputations, and undermines trust" (descriptive). "We should defend democracy from totalitarianism" (prescriptive) is another way of saying "Democracy is far more conducive to human flourishing than the alternatives are" (descriptive). . . . Imagine that you could push a button that would make every person on earth a little more creative, compassionate, intelligent, and fulfilled—in such a way as to produce no negative effects, now or in the future. This would be "good" in the only moral sense of the word that I understand. However, to make this claim, one needs to posit a larger space of possible experiences (e.g., a moral landscape). What does it mean to say that a person *should* push this button? It means that making this choice would do a lot of good in the world without doing any harm. And a disposition to not push the button would say something very unflattering about him. After all, what possible motive could a person have for declining to increase everyone's well-being (including his own) at no cost? I think our notions of "should" and "ought" can be derived from these facts and others like them. Pushing the button is better for everyone involved. What more do we need to motivate prescriptive judgments like "should" and "ought"? (Harris 2014)

Harris's partner-in-scientism Michael Shermer attempts to cross the is/ought boundary in a similar way:

> Morality involves how we think and act toward other moral agents in terms of whether our thoughts and actions are right or wrong with regard to their *survival* and *flourishing*. By *survival* I mean the instinct to live, and

by *flourishing* I mean having adequate sustenance, safety, shelter, bonding, and social relations for physical and mental health. Any organism subject to natural selection—which includes all organisms on this planet and most likely on any other planet as well—will by necessity have this drive to survive and flourish, for if they didn't they would not live long enough to reproduce and would therefore no longer be subject to natural selection. . . . Given these reasons and this evidence, the *survival and flourishing* of sentient beings is my starting point, and the fundamental principle of this system of morality. It is a system based on science and reason, and is grounded in principles that are themselves based on nature's laws and on human nature—principles that can be tested in both the laboratory and in the real world. (Shermer 2015, 11–12)[2]

Many others[3] have pointed out the numerous moral and evaluative assumptions Harris and Shermer must rely on in order for their approach to work. My aim here is not to repeat these important criticisms but rather to point out a separate but very general difficulty that faces any proponent of scientism who attempts to go this route. For what Shermer, Harris, and others have in common is their attempt to cross the is/ought gap by tucking the real work into *semantics*. Their arguments can work only if we assume evaluative definitions of quasi-naturalistic terms like *survival* and *flourishing*. In that case, it would be those definitions, and not the arguments, that do the heavy lifting. Is this a legitimate solution to the problem? While it would be difficult to refute all such attempts categorically, the prospects seem extremely bleak for scientism. For—and this is the crucial point—scientism is not only committed to ethical naturalism. It is committed, much more significantly, to the broader and more extreme view that *all* facts, including not only ethical facts but also philosophical facts more generally (including semantic facts), must ultimately be observational, logical, mathematical, or scientific facts in a nontrivial sense of *scientific*.

Once this has been understood, the prospects for scientism in ethics appear to be far dimmer than the prospects of merely bridging the is/ought divide by appropriately defining one's logical operators or converting normative theory problems into semantics problems. Perhaps Shermer and Harris hold that questions about the proper semantics of logical operators or ordinary language can be resolved by appeals to common usage, hence making semantic facts a subset of social science facts. But how would this work in the case of logical operators? And if the claim that morality is whatever promotes survival and flourishing of sentient beings is meant to be another way of making a statistical prediction about what English speakers would assent to, then there are in fact two separate difficulties to face. The first is that of squaring this empirical matter with what we are after when we try to determine what is morally right or wrong; the second is that the empirical predictions that seem to be implied about survey results seem unlikely to turn out in Harris's and Shermer's favor.

To see this clearly, it is useful to consider how Shermer and Harris seem to have arrived at their very odd definitions of key terms. If they were to say that acting morally were *by definition* identical with whatever leads to the maximum survival of members of our species, or perhaps of the majority of its members (using *survival* as *most* competent English speakers understand it), and if we were also to use the linguistic intuitions of English speakers as the criterion that determines the correct definition of a term, then the moral claims they are making are empirical but very likely false. Most English speakers, one suspects, would not agree with statements like "In the event of a nuclear holocaust that leaves the human race irreparably genetically altered in such a way that the current generation and all future generations will consist entirely of individuals in constant and excruciating agony, seldom capable of thought (and then only at the lowest level), and uninterested in reproduction, it is morally necessary to force humans to keep reproducing for the longest possible time." If acting morally were instead defined as whatever leads to the maximum survival of conscious or self-conscious beings regardless of species, then the same counterexample applies if we add the condition that there are no other conscious beings in the universe. This is why, in order to make their views even slightly plausible, Shermer and Harris are both obliged to add the evaluative (and conveniently slippery) term *flourishing* to the nonevaluative *survival*. It is also, presumably, part of why Shermer parts ways with dictionaries and defines *survival* not in terms of actually continuing to exist but rather in terms of the *instinct that impels* people to *wish* to continue existing.

Shermer must do much more than simply define his key terms plausibly and naturalistically, though: for his project to work, he must also build moral principles out of them without adding in any moral assumptions. It is here that his difficulties become particularly revealing. What he offers as a moral "principle" is clearly nothing of the sort. Again, what he tells us is that "the *survival and flourishing* of sentient beings is my starting point, and the fundamental principle of this system of morality. It is a system based on science and reason, and is grounded in principles that are themselves based on nature's laws and on human nature—principles that can be tested in both the laboratory and in the real world." But "the survival and flourishing of sentient beings" cannot be a "principle" in the moral or even the scientific sense of the term: it doesn't tell us what to do or what to expect. Whatever real work Shermer does to get himself from an is to an ought is hidden elsewhere, and the reference to testing in "the laboratory" is merely a convenient scientific-sounding cover, devoid of content, whose purpose seems to be that of giving credibility to the apparently impossible enterprise without giving his audience a glimpse of how this is meant to work in practical terms. It is surprising that such a committed opponent of pseudoscience in its other forms would fail to recognize its hallmark features here.

Harris does slightly better than Shermer in making clear how he intends to bridge the gap, as we have seen. He says, for instance, that "'You shouldn't lie' (prescriptive) is synonymous with 'Lying needlessly complicates people's lives, destroys reputations, and undermines trust' (descriptive)." In saying this, Harris owes us a plausible account of synonymy according to which this and other such claims come out true—but then it is not clear how *that* account could be arrived at in a genuinely scientific manner. Also, surely things cannot be as simple as Harris claims even in this limited case, since we could by the same token say that the implausible prescriptive statement "You shouldn't expose the diabolical plot of this corrupt public figure" is synonymous with the plausible descriptive statement "Exposing the diabolical plot of this corrupt public figure would needlessly complicate people's lives, destroy reputations, and undermine trust."

Harris does of course provide himself with the weasel word *needlessly* here, but now the question becomes, what counts as needless, and how does one determine *this* in a purely scientific manner? The problem is now not only that Harris needs to derive an ought statement from is statements, but also that he has to balance the resulting, defeasible ought statement against many other defeasible ought statements, and (for his project to succeed in an interesting way) he must make these comparative judgments fully within the confines of empirical science. Later in the same quoted passage, Harris adverts to the *disposition* and *motive* of a potential button pusher in support of a hypothetical moral verdict. Since Harris is arguing that morality follows logically from some empirical facts plus the correct definitions of moral and related terms, then surely he owes us an account of what those correct definitions are *and how he knows that they are the correct definitions*. So he has two formidable challenges here, not one. He first has to come up with a clear, nonevasive standard by which to bridge the is/ought gap ("An act is moral if and only if it has property X"). This is the part of the problem that involves the threat of the so-called naturalistic fallacy, since one could (as Moore says) respond by saying, "I understand that it has property X; but it's not yet clear to me that it must be moral." But Harris also faces another challenge, which is to articulate and defend accounts of synonymy and correct definitions that are both plausible and *wholly* a matter of empirical science, logic, mathematics, and observation. And about this, we have so far heard nothing.

To summarize the discussion so far: there may not be a general basis for thinking that the is/ought gap is unbridgeable in principle, for the reasons Pigden discusses. But scientism has a more serious, and more general, problem to deal with in the naturalization of ethics: the problem not of deriving an ought from an is but rather of deriving the philosophical from the non-philosophical. If the is/ought gap can be crossed legitimately, the crossing will be possible (and its legitimacy defended) only by means of some careful semantic or other maneuvering. But each of these maneuvers incurs a further cost, which can be paid off only in the currency of philosophy; and to naturalize that part of philosophy also, one must take on yet further costs. While this need not be a problem at all for Pigden and other philosophers who merely wish to derive an ought from an is, it is a significant problem for Shermer, Harris, and other devotees of scientism who are perpetually stuck borrowing large sums at high interest rates from one area of philosophy to pay off smaller debts incurred in other areas of philosophy, all in the name of freeing themselves of philosophical debt entirely.[4] If there are serious prospects for scientism in ethics, any case for those prospects had better address this problem, which is more general and pressing than the limited problem of the naturalistic fallacy in ethics.

Still, while Harris, Shermer, and other advocates of scientism are far from successful in their attempts to discharge their argumentative burdens, it seems a step too far to dismiss outright some other scientific challenges to substantive moral views. Jonathan Haidt (2012) and Joshua Greene (2013) are two prominent examples of recent thinkers who have brought the tools of empirical investigation into the moral sphere[5] with more philosophical sophistication than Harris or Shermer have shown. Greene, indeed, is a former student of the renowned utilitarian Peter Singer. Both Haidt and Greene present their readers with unsettling evidence that our moral judgments are affected much more by emotional influences than most of us find it natural to assume.

One example of a nonrational influence on our moral judgments that Haidt discusses is the phenomenon of moral dumbfounding. In a series of experiments, Haidt and others present various subjects with disturbing moral scenarios: a family member accidentally runs over the household pet in a car, and the family then cooks and eats the pet; an adult brother and sister both willingly decide to try having sexual intercourse just once, using several birth control methods, and then tell no one; someone regularly has sex with animal carcasses purchased as meat; another character tears up a flag and uses the strips to clean the toilet in her private home; and so on. These vignettes are carefully constructed to allow for no easy explanation as to why exactly the situations they depict are immoral, but most subjects in Haidt's experiments tend to find them morally troubling and condemn the characters' actions unequivocally. It is at this point that the interesting part of the experiment begins. The subjects are pressed to justify their negative moral judgments, and find it difficult to do so. The experimenters are trained to respond to all the most likely moral justifications with counterarguments showing that the subject would be inconsistent if he or she relied on the same moral reasoning elsewhere. This produces the so-called moral dumbfounding Haidt is so interested in: the strong sense that something is morally wrong even though one seems unable to explain why. Faced with the strong replies to their attempts to ground their moral intuitions in plausible moral principles, Haidt's subjects must choose between abandoning their judgments and maintaining them in the absence of any adequate support. They tend strongly to choose the latter course, rather unlike the process we seem to follow in cases where we learn of information that undermines key premises we have used in support of some view. This, Haidt argues, provides evidence against the view that we arrive at moral conclusions on the basis of abstract reasoning from moral principles.

Greene, like Haidt, sees our capacity for moral judgment as bearing clear marks of evolutionary adaptation. On Greene's view, many of our most common emotional tendencies—moral disgust, desire for even self-destructive vengeance when we have been seriously betrayed, awe, loyalty, and gossip—are best explained as arising from a successful set of evolutionary "strategies" for getting us to cooperate rather than defect in prisoner's dilemma cases. To take one example of Greene's reasoning here: a member of a species of perfectly rational and purely self-interested beings would be likely not only to defect in a prisoner's dilemma case but also to make dishonest promises in advance about cooperating in such cases, so long as he or she recognizes that it is disadvantageous for the other party to take revenge. But since vengeance is typically costly, and members of a perfectly rational species would understand that other members of the same species would recognize both this and the fact that everyone else would recognize it, the empty threat of revenge would not be very effective at motivating them to make and keep promises to cooperate. But the threat of vengeance would be very compelling if the beings were aware of universally powerful urges that lead members of the species to avenge themselves for betrayals even at the cost of a significant personal sacrifice. Greene then seeks to verify these game-theoretical explanations of morality and examine how they play out neurologically by presenting subjects in fMRIs with familiar moral dilemmas like the trolley problem and comparing their spoken solutions with their brain activity. In the end, Greene presses for agnosticism on what he calls the "deeper" question of what makes certain actions objectively moral, but makes a case for adopting utilitarianism in cases of entrenched moral conflict as a pragmatic means of conflict resolution.

Not all empirical investigators of our ethical tendencies seek to explain our moral judgments in evolutionary terms. Marvin Harris (1975; 1998), Jesse Prinz (2007), and Richard Nisbett and Dov Cohen (1996) have offered plausible cultural and environmental explanations of a wide range of moral attitudes and practices, ranging from dietary restrictions to the localized tendency to react violently against insults to the taboo against cannibalism.

Read in what I see as a plausibly charitable light,[6] Haidt, Greene, Marvin Harris, Prinz, Nisbett, Cohen, and others—perhaps even Shermer and Sam Harris, at their better moments—are engaged in a less radical project than out-and-out scientism. Rather than suggesting that scientific research into moral judgments and practices eliminates the need for philosophical reflection in normative reasoning, they present empirical evidence that challenges our commonsense moral picture in ways that can undermine our confidence in some of our deepest philosophical intuitions. It seems to sharpen, rather than eliminate, the need to reflect philosophically on what follows.

One way of replying to those who employ scientistic thinking in this weaker sense is to contest some presented finding or their interpretation of it. For instance, Kahane and colleagues (2014) have recently argued that experiments used by Greene and others to support various substantive conclusions about morality actually misidentify other, perhaps *psychopathic*, tendencies as utilitarian ones. While such responses can cast serious doubt on certain weakly scientific arguments about morality by beating those who present them at their own game (Kahane et al. argue against the original findings by conducting a study of their own), they at best respond to *particular* moves in the game of empirical moral research. They cannot stand as *categorical* objections against empirical moral research as a whole, since they are a part of that very research.

To rule out a priori the relevance of all empirical research and evolutionary reasoning to substantive morality, we need a general philosophical argument. Massimo Pigliucci provides such an argument in a recent response to Greene:

> Let me interject here with my favorite example of why exactly Greene's reasoning doesn't hold up: mathematics. Imagine we subjected a number of individuals to fMRI scanning of their brain activity while they are in the process of tackling mathematical problems. I am positive that we would find the following (indeed, for all I know, someone might have done this already):
>
> There are certain areas, and not others, of the brain that lit up when someone is engaged with a mathematical problem.
>
> There is probably variation in the human population for the level of activity, and possibly even the size or micro-anatomy, of these areas.
>
> There is some sort of evolutionary psychological story that can be told for why the ability to carry out simple mathematical or abstract reasoning may have been adaptive in the Pleistocene (though it would be much harder to come up with a similar story that justifies the ability of some people to understand advanced math, or to solve Fermat's Last Theorem).
>
> But none of the above will tell us anything at all about whether the people in the experiment got the math right. Only a mathematician—not a neuroscientist, not an evolutionary psychologist—can tell us that. (Pigliucci 2016)

Pigliucci is surely right to point out that the justification of a moral judgment is very different from its causal explanation, and that a moral theory cannot be straightforwardly vindicated by a discovery that that theory is common, innate, or even game-theoretically optimal unless one has an accompanying evaluative or normative premise.[7] And in an important sense, mathematical reasoning may well stand in the same relation to mathematics that moral thinking stands in relation to morality. But the success of Pigliucci's interesting analogy in ruling out Greene's project *or anything along similar lines* depends on how the mathematical analogy accommodates what Greene, Haidt, and others would presumably see as significant differences between the two. Among those apparent differences are these: there is simply far greater intercultural and intracultural disagreement about morality than there is about mathematics; it is at least much more difficult to verify simple moral claims objectively than it is to find objective support for the claims of basic arithmetic; and mathematical judgments tend to survive emotional manipulation unchanged in ways that moral judgments do not. It seems that Pigliucci's argument by analogy could be defended in one of two ways.

The first strategy would be to argue that the empirical differences between the psychology of mathematical judgment and the psychology of moral judgment are not really so important. Perhaps there is a good case to be made that there is as much evidence of disagreement on mathematical issues as there is on moral issues, or that, where moral and mathematical judgments of the same level of sophistication are made, the evidence of emotional influence turns out to be equally strong or weak, and so on. The problem with this approach is that even if it is successful, it gives the larger game away by engaging the debate on empirical terms, thus implicitly conceding the relevance of empirical issues to moral ones.

The second strategy is the one taken by Pigliucci himself in a brief, private conversation with me on the subject.[8] This involves seeing the study of morality as a completely abstract enterprise on a par with pure mathematics or geometry. There is no purely mathematical reason to opt for non-Euclidian over Euclidian geometry, and the theorems we prove in Euclidian geometry are equally well supported whether or not Euclidian geometry turns out to hold in the real world. Similarly, we could take morality to be a domain of inquiry that is, *by definition*, completely closed to empirical support or refutation. We could then fashion any number of moral systems, each with its own set of axioms, and it would be trivially true that any attempt to evaluate those moral systems on the basis of experimentation or evolutionary theorizing would commit a category mistake. The difficulty for this strategy is that we generally want to know, in the actual world, which actions are morally permissible and which are not. If the moralist can only point to an unlimited range of moral systems and say of some action that it is permitted in these but not those, and that we have no way of knowing which moral system (if any) applies to our world, then it is hard to see that the project of ethical inquiry is very helpful. Conversely, if there are signs in our world that indicate which a priori moral theory applies, akin to the scientific evidence that shows Euclidian geometry not to apply to our world, then the door is open again to the relevance of empirical findings in practical moral thinking.

It is also possible to take neither an empirical nor an antiempirical strategy in defending Pigliucci's analogy with mathematics, and to insist instead that the lines of evidence brought by Haidt, Greene, and others against the reliability of our moral intuitions simply don't matter to the content of morality, since the project of moral inquiry has to do with what *is* moral, not with the accuracy or trustworthiness of our *judgments* about what is moral. But this approach has its own drawbacks. To continue the analogy with mathematics: if we were to discover that people in one psychological state can reliably be predicted to make mathematical judgments that are noticeably and significantly different from those who are not in that psychological state, and if there were no clear, objective way to compare the judgments of the two groups with some mind-independent facts to see whether the psychological state enhanced or hindered mathematical judgments, then it would surely be unwise to remain as confident as before in our ability to do mathematics objectively. Similarly, if the empirical evidence that Haidt, Greene, and others present for the unreliability of our moral judgments came to be supplemented with further empirical evidence that is *a thousand times stronger*, and to affect *all* our moral judgments, then it would be unreasonable for us to maintain confidence in our moral theories. Perhaps a general agnosticism about morality would be the most reasonable course, then; and this epistemological position could in turn be used to support a metaphysical theory of subjectivism, error theory, and so on, thereby supporting some currently implausible views in normative theory and applied ethics. But the view that empirical arguments along the lines of Haidt's and Greene's should be discounted immediately on the grounds that they attempt to bring empirical issues into a nonempirical discussion implies that such empirical findings would continue to be irrelevant even if they were a thousand, or even a million, times stronger.

To sum up: the crude attempts of Shermer and Sam Harris to *completely replace* philosophical reasoning with scientific investigations in morality seem hopeless, since that project implicitly relies on at least as many philosophical assumptions as it seeks to replace; moreover, the philosophical assumptions the project relies on are highly dubious for familiar reasons. But there is room for many more moderate empirical projects in ethics that do not originate from a naïve misunderstanding of the philosophical enterprise; and it is possible for those projects to legitimately call into question not only the application of normative ethical theories to particular cases but also our basis for believing in some substantive normative and metaethical theories. Rather than dismiss all such projects out of hand, philosophers ought to take them seriously and accept or reject them on a case-by-case basis.

......

Notes to Chapter Five

1. Pigden is referring here to Schurz 1997.

2. https://www.academia.edu/5664257/The_Is-Ought_Gap, p. 8, accessed February 20, 2017.

3. See, e.g., Earp 2011 and Pigliucci 2013. And see Blackford 2010, which also comments briefly on the implausible moral semantics in *The Moral Landscape*.

4. In case one of my reviewers is not alone in finding this metaphor opaque: the scientists here are claiming that they can rid their moral conversations of all debts to philosophy, but are able to make good on their promise only by relying on various assumptions in semantics and metaphysics that will have to be "paid off" with philosophical arguments that never seem to be produced.

5. For a discussion of their general approach and *its* scientist tendencies, see Sorell in this volume.

6. Here is an example of what I mean by reading in a charitable light: John Gray (2014) makes much of the following quote by Greene: "Morality is a set of psychological adaptations that allow otherwise selfish individuals to reap the benefits of social cooperation." Does Greene really feel that morality *itself* is a set of psychological adaptations, Gray wonders? (It is difficult to know what that could mean, since any faculty we use for thinking about, reacting to, learning, or even creating facts of some sort cannot, presumably, be identical with those facts themselves.) Or is this a brief, albeit misleading, way of saying that our *capacity for moral judgment* is a set of psychological adaptations? (That would be true, but almost trivially so.) I read Greene as saying here something parallel to "The human capacity for mathematical thinking is a set of psychological adaptations that allow individuals to reap the benefits of precise calculations," neither explicitly rejecting nor accepting the existence of a mind-independent component to the domain. If I am mistaken in my interpretation of Greene, my attempt to read him charitably may at least point to a consistent alternative position on the issue. I propose that it will be useful for his critics to read him *as if* he were adopting this position. The important issue, after all, is not whether Haidt and Greene are out of their element in discussing morality: it is whether a range of ethical and metaethical views many of us hold are undermined by certain empirical considerations, and more broadly, whether such empirical undermining of core ethical or metaethical views is possible.

7. I should state here that I am not at all convinced that Greene is in fact making any of these errors. But I leave that issue aside, since my purpose here is to examine whether it is legitimate to rule out *any possible* project along the lines of Greene's, not to assess the merits of Greene's project itself.

8. This was in the early summer of 2014.

Literature Cited

Blackford, Russell. 2010. "Book Review: Sam Harris' *The Moral Landscape*." *Journal of Evolution and Technology* 21, no. 2 (December): 53–62.

Earp, Brian. 2011. "Sam Harris Is Wrong about Science and Morality." *Practical Ethics*, November 17, http://blog.practicalethics.ox.ac.uk/2011/11/sam-harris-is-wrong-about-science-and-morality/.

Gray, John. 2014. "*Moral Tribes: Emotion, Reason, and the Gap between Us and Them* by Joshua Greene—review: Is This Call for Rational Thinking to Resolve Major Conflicts Crude Reductionism?" *Guardian* (Manchester), January 17, http://www.theguardian.com/books/2014/jan/17/moral-tribes-joshua-greene-review.

Greene, Joshua D. 2013. *Moral Tribes: Emotion, Reason, and the Gap between Us and Them.* New York: Penguin.

Haidt, Jonathan. 2012. *The Righteous Mind: Why Good People Are Divided by Politics and Religion.* New York: Pantheon.

Harris, Marvin. 1975. *Cows, Pigs, Wars and Witches: The Riddles of Culture.* London: Hutchinson.

———. 1998. *Good to Eat: Riddles of Food and Culture.* Long Grove, IL: Waveland Press.

Harris, Sam. 2005. *The End of Faith.* New York: W. W. Norton.

———. 2010. "Moral Confusion in the Name of Science," samharris.org (blog), http://www.samharris.org/blog/item/moral-confusion-in-the-name-of-science/.

———. 2014. "Clarifying the Landscape," samharris.org (blog), https://www.samharris.org/blog/item/clarifying-the-landscape.

Kahane, Guy, Jim A. C. Everett, Brian D. Earp, Miguel Farias, and Julian Savulescu. 2014. "'Utilitarian' Judgments in Sacrificial Moral Dilemmas Do Not Reflect Impartial Concern for the Greater Good." *Cognition* 134 (January): 193–209.

Nisbett, R. E., and D. Cohen. 1996. *Culture of Honor: The Psychology of Violence in the South.* Boulder, CO: Westview.

Pigden, Charles. 2013. "The Is-Ought Gap." In Lafollette et al., *The International Encyclopedia of Ethics*, pp. 11–13. Oxford: Wiley-Blackwell. Article viewed online at https://www.academia.edu/5664257/The_Is-Ought_Gap.

Pigliucci, Massimo. 2013. "Michael Shermer on Morality." *Rationally Speaking* (blog), January 21, http://rationallyspeaking.blogspot.com/2013/01/michael-shermer-on-morality.html.

———. 2016. "The Problem with Cognitive and Moral Psychology." *The Philosophers' Magazine Online*, February 10, https://platofootnote.wordpress.com/2016/02/08/the-problem-with-cognitive-and-moral-psychology/.

Prinz, Jesse. 2007. *The Emotional Construction of Morals.* Oxford: Oxford University Press.

Schurz, Gerhard. 1997. *The Is-Ought Problem: A Study in Philosophical Logic.* Dordrecht, Netherlands: Kluwer.

6

· · · · · · · · · ·

The Trouble with Scientism:
Why History and the Humanities Are Also a
Form of Knowledge

PHILIP KITCHER

There are two cathedrals in Coventry. The newer one, consecrated on May 25, 1962, stands beside the remains of the older one, which dates from the fourteenth century, a ruin testifying to the bombardment of the Blitz. Three years before the consecration, in one of the earliest ventures in the twinning of towns, Coventry had paired itself with Dresden. That gesture of reconciliation was recapitulated in 1962, when Benjamin Britten's *War Requiem* received its first performance at the ceremony. The three soloists were originally intended to come from three nations that had suffered heavy casualties in the two world wars. There was to be an English tenor (Peter Pears), a German baritone (Dietrich Fischer-Dieskau), and a Russian soprano (Galina Vishnevskaya). But the Soviet Union denied a visa to Vishnevskaya, and she was replaced by Heather Harper.

Since the 1960s, historians have worked—and debated—to bring into focus the events of the night of February 13, 1945, in which an Allied bombing devastated the strategically irrelevant city of Dresden. An increased understanding of the decisions that led to the firebombing, and of the composition of the Dresden population that suffered the consequences, have altered sub-

Originally published in the *New Republic*, May 24, 2012.

sequent judgments about the conduct of war. The critical light of history has been reflected in the contributions of novelists and critics, and of theorists of human rights. Social and political changes, in other words, followed the results of humanistic inquiry, and were intertwined with the reconciliatory efforts of the citizens of Coventry and Dresden. Even music and poetry played roles in this process: what history has taught us is reinforced by the lines from the English soldier and poet Wilfred Owen that Britten chose as the epigraph for his score—"My subject is war, and the pity of war. The poetry is in the pity. All a poet can do today is warn."

It is so easy to underrate the impact of the humanities and of the arts. Too many people, some of whom should know better, do it all the time. But understanding why the natural sciences are regarded as the gold standard for human knowledge is not hard. When molecular biologists are able to insert fragments of DNA into bacteria and turn the organisms into factories for churning out medically valuable substances, and when fundamental physics can predict the results of experiments with a precision comparable to measuring the distance across North America to within the thickness of a human hair, their achievements compel respect, and even awe. To derive our notion of human knowledge from the most striking accomplishments of the natural sciences easily generates a conviction that other forms of inquiry simply do not measure up. Their accomplishments can come to seem inferior, even worthless, at least until the day when these domains are absorbed within the scope of "real science."

The conflict between the *Naturwissenschaften* and the *Geisteswissenschaften* goes back at least two centuries, and intensified as ambitious, sometimes impatient researchers proposed to introduce natural scientific concepts and methods into the study of human psychology and human social behavior. Their efforts, and the attitudes of unconcealed disdain that often inspired them, prompted a reaction, from Vico to Dilthey and into our own time: the insistence that some questions are beyond the scope of natural scientific inquiry, too large, too complex, too imprecise, and too important to be addressed by blundering oversimplifications. From the nineteenth-century ventures in mechanistic psychology to contemporary attempts to introduce evolutionary concepts into the social sciences, scientism has been criticized for its "mutilation" (*Verstümmelung*, in Dilthey's memorable term) of the phenomena to be explained.

The problem with scientism—which is of course not the same thing as science—is owed to a number of sources, and they deserve critical scrutiny.

The enthusiasm for natural scientific imperialism rests on five observations. First, there is the sense that the humanities and social sciences are doomed to deliver a seemingly directionless sequence of theories and explanations, with no promise of additive progress. Second, there is the contrasting record of extraordinary success in some areas of natural science. Third, there is the explicit articulation of technique and method in the natural sciences, which fosters the conviction that natural scientists are able to acquire and combine evidence in particularly rigorous ways. Fourth, there is the perception that humanists and social scientists are able to reason cogently only when they confine themselves to conclusions of limited generality: insofar as they aim at significant—general—conclusions, their methods and their evidence are unrigorous. Finally, there is the commonplace perception that the humanities and social sciences have been dominated, for long periods of their histories, by spectacularly false theories, grand doctrines that enjoy enormous popularity until fashion changes as their glaring shortcomings are disclosed.

These familiar observations have the unfortunate effect of transforming differences of degree into differences of kind, as enthusiasts for the alleged superiority of natural science readily succumb to stereotypes and overgeneralizations without regard for more subtle explanations. Let us consider the five foundations of this mistake in order.

The most obvious explanation for the difficulties of the *Geisteswissenschaften*, the humanities and the study of history and society, is that they deal with highly complex systems. Concrete results are often achieved in particular instances: historians and anthropologists are able to be precise and accurate by sacrificing generality, by clearheadedly disavowing the attempt to provide any grand overarching theory. No large vision of history emerges from our clearer understanding of the bombing of Dresden, but the details are no less powerful and significant. In this respect, moreover, matters are no different in the natural sciences. As we shall see, science often foregoes generality to achieve a precise and accurate answer to an important question.

In English we speak about science in the singular, but both French and German wisely retain the plural. The enterprises that we lump together are remarkably various in their methods, and also in the extent of their successes. The achievements of molecular engineering or of measurements derived from quantum theory do not hold across all of biology, or chemistry, or even physics. Geophysicists struggle to arrive at precise predictions of the risks of earthquakes in particular localities and regions. The difficulties of intervention and prediction are even more vivid in the case of contemporary climate science:

although it should be uncontroversial that Earth's mean temperature is increasing, and that the warming trend is caused by human activities, and that a lower bound for the rise in temperature by 2200 (even if immediate action is taken) is two degrees Celsius, and that the frequency of extreme weather events will continue to rise, climatology can still issue no accurate predictions about the full range of effects on the various regions of the world. Numerous factors influence the interaction of the modifications of climate with patterns of wind and weather, and this complicates enormously the prediction of which regions will suffer drought, which agricultural sites will be disrupted, what new patterns of disease transmission will emerge, and a lot of other potential consequences about which we might want advance knowledge. (The most successful sciences are those lucky enough to study systems that are relatively simple and orderly. James Clerk Maxwell rightly commented that Galileo would not have redirected the physics of motion if he had begun with turbulence rather than with free fall in a vacuum.)

The emphasis on generality inspires scientific imperialism, conjuring a vision of a completely unified future science encapsulated in a "theory of everything." Organisms are aggregates of cells, cells are dynamic molecular systems, the molecules are composed of atoms, which in their turn decompose into fermions and bosons (or maybe into quarks or even strings). From these facts it is tempting to infer that all phenomena—including human actions and interactions—can "in principle" be understood ultimately in the language of physics, although for the moment we might settle for biology or neuroscience. This is a great temptation. We should resist it. Even if a process is constituted by the movements of a large number of constituent parts, this does not mean that it can be adequately explained by tracing those motions.

A tale from the history of human biology brings out the point. John Arbuthnot, an eighteenth-century British physician, noted a fact that greatly surprised him. Studying the registry of births in London between 1629 and 1710, he found that all the years he reviewed showed a preponderance of male births: in his terms, each year was a "male year." If you were a mad devotee of mechanistic analysis, you might think of explaining this—"in principle"—by tracing the motions of individual cells, first sperm and eggs, then parts of growing embryos, and showing how the maleness of each year was produced. But there is a better explanation, one that shows the record to be no accident. Evolutionary theory predicts that for many, but not all, species, the equilibrium sex ratio will be 1:1 at sexual maturity. If it deviates, natural selection will favor the underrepresented sex: if boys are less common, invest in sons and you are likely to have more grandchildren. This means that if one sex is more likely to die before reaching reproductive age, more of that sex will have to be produced to start with. Since human males are the weaker sex—that is, they are more likely to die between birth and puberty—reproduction is biased in their favor.

The idea of a "theory of everything" is an absurd fantasy. Successful sciences are collections of models of different types of phenomena within their domains. The lucky ones can generate models that meet three desiderata: they are general, they are precise, they are accurate. Lots of sciences, natural sciences, are not so fortunate. As the ecologist Richard Levins pointed out decades ago, in many areas of biology—and, he might have added, in parts of physics, chemistry, and earth and atmospheric science as well—the good news is that you can satisfy any two of these desiderata, but at the cost of sacrificing the third. Contemporary climatology often settles for generality and accuracy without precision; ecologists focusing on particular species provide precise and accurate models that prove hard to generalize; and of course if you abandon accuracy, precision and generality are no problem at all.

Let us turn now to the celebration of scientific rigor. Individual sciences develop rules and standards for appraising evidence—as they learn about aspects of nature, they learn more about how to learn. At any particular stage of inquiry, communities of scientists agree on the canons of good inference, so that the work of certification of new results goes relatively smoothly. To the extent that the agreed-on rules are reliable, knowledge accumulates. It is important to understand, however, that at times of major change the standards of good science themselves are subject to question and discussion. And this observation, amply demonstrated in the history of the sciences, has important consequences.

The contrast between the methods of the two realms, which seems so damning to the humanities, is a false one. Not only are the methods deployed within humanistic domains—say, in attributions of musical scores to particular composers or of pictures to particular artists—as sophisticated and rigorous as the techniques deployed by paleontologists or biochemists, but in many instances they are the same. The historical linguists who recognize connections among languages or within a language at different times, and the religious scholars who point out the affiliations among different texts, use methods equivalent to those that have been deployed ever since Darwin in the study of the history of life. Indeed, Darwin's paleontology borrowed the method from early nineteenth-century studies of the history of languages.

By the same token, the sense that the humanities are dominated by changes in fashion among spectacularly false theories suggests a contrast where none is to be found. If Marx and Freud are favorite whipping boys for those worried about the *Geisteswissenschaften*, the compliment is readily returned. Not only did behaviorist psychology—itself motivated by the desire to make studies of human conduct "truly scientific"—dominate much of twentieth-century social science, but its influence was foreshadowed within nineteenth-century physics and chemistry with the proliferation of "ether theories." No less a figure than Maxwell even characterized the ether as "the best confirmed entity in natural philosophy" (by which he, like his contemporaries, meant "natural science").

So the five points of scientism rest on stereotypes, and these are reinforced by the perception of threats. As the budgets for humanities departments shrink, humanists see natural scientists blundering where the truly wise fear to tread. Conversely, scientists whose projects fail to win public approval seem to envision what John Dupré has called the "Attack of the Fifty-Foot Humanist," a fantasy akin to supposing that postmodernist manifestos are routinely distributed with government briefing books. We need to move beyond the stereotypes and discard the absurd visions that often maintain them.

To develop a better view, let us focus for the moment on history and anthropology (although the conclusions I shall defend would also apply to literature and the arts, as well as to critical studies of them). While it is true that rigorous history and ethnography often give up generality for accuracy and precision, their conclusions can nonetheless have considerable importance. Scientific significance is not limited to the discovery of general laws—that idea is a hangover from an age in which the scientific task was seen as one of fathoming the Creator's rulebook, of thinking "God's thoughts after Him." The sciences, recall, are collections of models, directed at answering questions. Not every question matters: there are countless issues about the variation of your physical environment while you read these sentences that should be of no concern to anybody. Generality is to be prized, partly because it is often the key to answering questions wholesale rather than retail, partly because generalizing explanations are often deeper; but there are many nongeneral issues, concrete and individual questions, that rightly occupy natural scientists. Where exactly do the fault lines run in Southern California? What is the relation among the various hominid species? By the same token, there are many specific questions that occupy historians and anthropologists.

Some of these questions are causal, about the factors that generated large events or that sustain particular social systems. Yet there are others that should be emphasized. When Emmanuel Le Roy Ladurie writes a study of a medieval population on the Franco-Spanish border, or when Jean L. Briggs reports on family life among the Inuit, these scholars are not primarily interested in tracing the causes of events. History and ethnography are used instead to show the readers what it is like to live in a particular way, to provide those of us who belong to very different societies with a vantage point from which to think about ourselves and our own arrangements. Their purpose, to borrow an old concept, is a kind of understanding that derives from imaginative identification.

Although studies such as these make no pretense at generality, their impact can be very large. They can unsettle the categories that are taken for granted in all kinds of decisions, from mundane reflections about how to respond to other people to large matters of social policy. "Collateral damage," for example, comes to seem an inappropriate way to talk about the victims of the Dresden firebombing. Humanistic studies can also challenge the categories used to frame lines of scientific inquiry. History and anthropology are sites at which new concepts are forged. Their deliverances can do what Thomas Kuhn memorably claimed for the study of the history of the sciences: they can change the images by which we are held. The Bush administration tacitly concurred with Kuhn's view when, at a time of shrinking budgets for the arts and humanities, it launched an initiative to support historical studies of iconic American figures and achievements. One effect of history (the verdicts of which Bush aimed to counter) may be a rethinking of social institutions. (I should add that neither in the natural sciences nor in human inquiry should we conclude that the applications tell the whole story of significance: comprehending something for its own sake also counts.)

Once the intertwining of human inquiry with social change has been recognized, it is easy to see why history and ethnography demand constant rewriting. Returning to the same materials is valuable when historians or anthropologists gain new evidence—like their colleagues in the natural sciences, they are sometimes lucky in acquiring new data, and thus led to revise. Yet there are other reasons for revisiting themes and episodes that have already been thoroughly treated. The history of the Roman Empire needs to be rewritten because the changes in our own society make new aspects of the past pertinent. Older histories may have played a useful role in generating styles of social thought that we take for granted, but in the light of our newer conceptions contemporary historians may view different questions as significant. This may leave the impression of an enterprise in which nothing ever accumulates, but the impression is incorrect. If Gibbon has been in many respects superseded, we should be nonetheless grateful for the impact that his monumental history made on his many readers. Historians return to Gibbon because his words are not ours—it would be odd to speak as he does of the "licentiousness," "prostitutions," and "chastity" of the empress Theodora. If our questions are different, it is because we live in a very different culture, one that his history helped to bring about.

The domain of the social sciences is the territory on which humanists and natural scientists frequently join battle. (It is also contested territory independent of this opposition: economists and sociologists offer competing accounts of the same phenomena, and of course there are conflicts within economics itself.) If the intertwining of history and ethnography with social thought and social change is appreciated, then the vehemence directed against natural scientific incursions into studies of human psychology and social behavior is more readily comprehensible. Since the conclusions reached in analyses of human behavior will be socially consequential—they may result in actual policies—the evidence for them deserves to be closely scrutinized, just as drugs that may have far-reaching side effects are subjected to the most careful testing.

To declare that there is a "natural unemployment rate" of 6 percent has a wide-ranging social and political impact, and it is entirely reasonable for critics to examine the evidence alleged to support such a declaration. Likewise, the outcry against early ventures in sociobiology was fueled by the perception that, while the claims advanced were sweeping (and sometimes threatening to the aspirations of large groups of people), the support for them was markedly less strong than that routinely demanded for theorizing about, say, insect sociality. In recent decades, proponents of Darwinian approaches to human behavior have been far more methodologically reflective, but the old war cries continue to echo in the contemporary context.

Human social behavior arises, in a complex social context, from the psychological dispositions of individuals. Those psychological dispositions are themselves shaped not only by underlying genotypes but also by the social and cultural environments in which people develop. Cultural transmission occurs in many animal species, but never to the extent or to the degree to which it is found in *Homo sapiens*. Human culture, moreover, is not obviously reducible to a complex system of processes in which single individuals affect others. Rigorous mathematical studies of gene-cultural coevolution reveal that when natural selection combines with cultural transmission, the outcomes reached may differ from those that would have been produced by natural selection acting alone, and that the cultural processes involved can be sustained under natural selection. Whether this happens in a wide variety of areas of human culture and domains or is relatively rare is something nobody can yet determine. But culture appears to be at some level autonomous and in some sense irreducible, and this is what scientism cannot grasp.

Let us imagine a model—call it the Big Model—of the causes of human social behavior. Natural selection would be a part of the story, but only a part. Work showing that particular social practices enhance reproductive success makes a welcome contribution, but more is needed if we are to draw conclusions about human practices as Darwinian adaptations. For if natural selection is at work, there must be a genetic basis for the psychological capacities and dispositions that underlie those practices. Moreover, the dispositions and the capacities are generated from the genetic basis in a particular environment or a range of environments, and it is important to understand how, in the pertinent environments, those genes give rise to other traits that might have quite different effects on reproductive success. And if genes, environments, and psychological characteristics are thoroughly entangled, it will be wrong to focus on forms of behavior that simply strike the investigator as potential adaptations. Moreover, the case analyzed must be one of those in which cultural transmission does not divert the outcome from that predicted by natural selection. (Even this is, strictly speaking, not enough, because it leaves unquestioned the assumption that the contribution of culture can be reduced to episodes in which items are transmitted from one individual to another.) All these lacks are difficult to make up in practice, because—as Darwin often lamented—"profound is our ignorance."

Beautiful and rigorous work in human behavioral ecology has been done, specifically in the construction of precise models and the measurement of contributions to reproductive success. But there is a tendency to inflate its conclusions, to think that the Darwinian part of the story is now in place and "cultural influences" can be left to mop up the rest. To draw such a conclusion is to overlook the fact that claims about adaptation presuppose a genetic basis for the relevant forms of behavior, and to assume that the case in question is one of those in which cultural transmission makes no difference. In effect, human behavioral ecologists simplify the Big Model in certain ways, and there is no advance reason to hold that what they have retained is more significant than what they have left out. Scientific analysis in this area would do better to proceed more symmetrically, to consider a variety of ways of simplifying an intractable Big Model, and to postpone firm conclusions about the operation of natural selection until a diversity of approaches has been explored. Welcome the precise studies, but propose conclusions tentatively. The political ramifications of conclusions about human beings only reinforce the demand for modesty.

For a variety of reasons, then, human inquiry needs a synthesis, in which history and anthropology and literature and art play their parts. But there is still a deeper reason for the enduring importance of the humanities. Many scientists and commentators on science have been led to view the sciences as a value-free zone, and it is easy to understand why. When the researcher enters the lab, many features of the social world seem to have been left behind. The day's work goes on without the need for confronting large questions about how human lives can or should go. Research is insulated because the lab is a purpose-built place, within which the rules of operation are relatively clear and well known. Yet on a broader view, which explores the purposes and their origins, it becomes clear that judgments of the significance of particular questions profoundly affect the work done and the environments in which it is done. Behind the complex and often strikingly successful practices of contemporary science stands a history of selecting specific aspects of the world for investigation. Bits of nature do not shout out "Examine me!" Throughout history, instead, innovative scientists have built a number of lampposts under which their successors can look. It is always worth considering whether the questions that now seem most significant demand looking elsewhere for new sources of illumination.

We are finite beings, and so our investigations have to be selective, and the broadest frameworks of today's science reflect the selections of the past. What we discover depends on the questions taken to be significant, and the selection of those questions, as well as the decision of which factors to set aside in seeking answers to them, presupposes judgments about what is valuable. Those are not only, or mainly, scientific judgments. In their turn, new discoveries modify the landscape in which further investigations will take place, and because what we learn affects how evidence is assessed, discovery shapes the evolution of our standards of evidence. Judgments of value thus pervade the environment in which scientific work is done. If they are made, as they should be, in light of the broadest and deepest reflections on human life and its possibilities, then good science depends on contributions from the humanities and the arts. Perhaps there is even a place for philosophy.

Healthy relationships between the sciences and the humanities should aspire to the condition of the best marriages—to a partnership in which different strengths and styles are acknowledged and appreciated, in which a fruitful division of labor constantly evolves, in which constructive criticism is given and received, in which neither party can ever make a plausible claim to absolute authority, and in which the ultimate goal is nothing less than the furtherance of the human good.

7

"Scientism!"

STEPHEN LAW

The term *scientism* is applied to a variety of positions about science. One is the view that the only legitimate questions about reality are those answerable by science. Another is that to the extent that anything can be known about reality, science alone is capable of providing that knowledge. Critics of religious, new age, spiritualist, and other, popular forms of divine or supernatural belief are often accused of scientism by their proponents. The accusation typically involves the thought that critics have crossed a line or boundary demarcating those topics or subjects that are the proper province of science, and those that are beyond its capacity to adjudicate. The accused are often found guilty of hubris, of an arrogant failure to recognize that there are "more things in heaven and Earth" than are dreamt of in their science, of supposing science is best placed to answer questions that, in reality, can be answered only by employing other disciplines, forms of inquiry, or "ways of knowing." Within discussion of religious, spiritual, new age, and popular divine or supernatural beliefs, this boundary marking the "limits of science" almost always plays an immunizing role: to explain why science constitutes no threat to such beliefs. "You scientists," say the believers, "may come *this* far, but no further."

Before we look at some specific examples of this sort of criticism, I want to outline some of my own views regarding the limits of science. As a philosopher, you might expect me to want to carve out some intellectual territory for philosophers to occupy, and also to resist the thought that philosophical questions and problems are either nonquestions and nonproblems or else questions that will be answered and problems that will be solved, if at all, through an application of the scientific method. I won't disappoint. However, while I acknowledge that there are limits to science, I will argue that these limits typically offer little comfort to religious, new age, and other folk looking for ways to immunize their beliefs against scientific refutation.

There is a further reason why I want to sketch out a case for supposing that philosophy is essentially a nonscientific enterprise: it will allow me to compare and contrast my own views with those of Richard Dawkins—who is perhaps Public Enemy Number One among those leveling the charge of scientism—and give me an opportunity to explain why Dawkins's position isn't as crudely scientistic as commonly supposed.

The Limits of Science

If science has limits, where do they lie? Below are two kinds of questions to which, it's very widely supposed, science can't supply answers.

1. Moral Questions

For those wishing to challenge the suggestion that science can answer every legitimate question, *moral* questions are often a first port of call. Hume drew our attention to the is/ought gap. Morality is concerned with what we *ought*, or *ought not*, to do. The empirical sciences, on the other hand, appear to be capable, in isolation, only of establishing what *is* the case. And it appears that premises concerning what is the case—certainly premises of the sort that pure empirical science is capable of establishing—fail rationally to support moral conclusions: conclusions about what we ought or ought not to do. But then it appears that science can't supply answers to our most fundamental moral questions. For science alone can neither directly reveal facts about what we ought or ought not to do, nor allow us legitimately to draw conclusions about what we ought or ought not to do.

Defenders of scientism respond in a variety of ways. Some suggest that there's something illegitimate about moral questions (that they're meaningless, perhaps). However, moral questions don't appear to be meaningless.

Indeed, we consider them among the most meaningful and pressing of questions. If they are ultimately illegitimate or meaningless, the onus is very much on the defenders of scientism wishing to take this route to show that that's the case. Alternatively, some defenders of scientism concede that while such questions are legitimate, and while science may not be able to answer them, there certainly isn't any *other* way of answering them. Another option is to deny there is any is/ought gap: to insist that the kind of facts science reveals *are* capable of justifying our most fundamental moral beliefs. Sam Harris (2012), in his book *The Moral Landscape*, recently developed a version of that view.

I won't take a view here regarding the is/ought gap, other than to note there's at least a prima facie case for supposing moral questions are questions science *alone* is ultimately incapable of answering. But of course, even if that conclusion is correct, it doesn't follow that science is morally irrelevant. Science remains capable of playing an important role in justifying and challenging many moral beliefs, most obviously those whose justification depends in part on empirical assumptions. If I believe that women ought not to have the vote because I believe both that people of low intelligence ought not to have the vote and that women are of low intelligence, then my justification can be straightforwardly shown to fail by scientific evidence that women are not of low intelligence. Similarly, if I believe it's morally right to enable human beings to flourish, then a scientific investigation into what actually enables humans to flourish becomes morally relevant. Science might also reveal that, say, at least some of our moral intuitions are not to be trusted by showing that what is actually shaping our intuitive moral responses is, in some cases, morally irrelevant. However, note that these suggestions concerning how science might play an important role in revealing moral truths and falsehoods are consistent with Hume's conclusion that science *alone* is incapable of justifying any moral position.

2. Philosophical Questions

On my view, philosophical questions are, for the most part, conceptual rather than scientific or empirical, and the methods of philosophy are, broadly speaking, conceptual rather than scientific or empirical.

Here's a simple conceptual puzzle. At a family get-together, the following relations held directly among those present: son, daughter, mother, father, aunt, uncle, niece, nephew, and cousin. Could there have been only four people present at that gathering? At first glance, there might seem to be a *conceptual obstacle* to there being just four people present—surely, more people are required for all those familial relations to hold among them? But in fact that appearance is deceptive. There could indeed be just four people present. To see that there being just four people present is not conceptually ruled out, we have to unpack, and explore the connections between, the various concepts involved. That is something that can be done from the comfort of your armchair.

Many philosophical puzzles have a similar character. Consider for example this puzzle associated with Heraclitus. If you jump into a river and then jump in again, the river will have changed in the interim. So it won't be the same. But if it's not the same river, then the number of rivers that you jump into is two, not one. It seems we're forced to accept the paradoxical—indeed, absurd—conclusion that you can't jump into one and the same river twice. Being forced into such a paradox by a seemingly cogent argument is a common philosophical predicament.

This particular puzzle is fairly easily solved: the paradoxical conclusion that the number of rivers jumped into is two, not one, is generated by a faulty inference. Philosophers distinguish at least two kinds of identity or sameness. Numerical identity holds where the number of objects is one, not two (as when we discover that Hesperus, the evening star, is identical with Phosphorus, the morning star). Qualitative identity holds where two objects share the same qualities (e.g. two billiard balls that are molecule-for-molecule duplicates of each other). We use the expression "the same" to refer to both sorts of identity. Having made this conceptual clarification, we can now see that the argument that generates our paradox trades on an ambiguity. It involves a slide from the true premise that the river jumped in the second time isn't qualitatively "the same" to the conclusion that it is not numerically "the same." We fail to spot the flaw in the reasoning because the words "the same" are used in each case. But now the paradox is resolved: we don't have to accept that absurd conclusion. Here's an example of how, by unpacking and clarifying concepts, it is possible to solve a classical philosophical puzzle. Perhaps not all philosophical puzzles can be solved by such means, but at least one can.

So some philosophical puzzles are essentially conceptual in nature, and some (well, one at least) can be solved by armchair, conceptual methods.

Still, I have begun with a simple, some might say trivial, philosophical example. What of the so-called hard problems of philosophy, such as the mind-body problem? The mind-body problem, or at least certain versions of it, also appears to be essentially conceptual in character. On the one hand, there appear to be reasons to think that if the mental is to have causal effects on the physical, then it will have to be identical with the physical. On the other hand, there appear to be conceptual obstacles to identifying the mental with the physical. Of course, scientists might establish various correlations between the mental and the physical. Suppose, for the sake of argument, that science establishes that whenever someone is in pain, his C-fibers are firing, and vice versa. Would scientists have then established that these properties are *one and the same property*—that pain just *is* C-fiber firing—in the way they have established that, say, heat just is molecular motion or water just is H_2O? Not necessarily. Correlation is not identity. It strikes many of us as intuitively obvious that pain just *couldn't be* a physical property like C-fiber firing—that these properties just *couldn't be* identical in that way. Of course, the intuition that certain things are ruled out can be deceptive. Earlier, we saw that the appearances of the concepts son, daughter, and so on are such that there just *had* to be more than four people at that family gathering was mistaken: when we unpack the concepts and explore the connections between them, it turns out there's no such conceptual obstacle. Philosophers have attempted to sharpen up the common intuition that there's an armchair obstacle to identifying pain with C-fiber firing or making some other physical property into a philosophical argument.

Consider Kripke's antiphysicalist argument, for example. Kripke's argument turns on the thought that the conceptual impossibility of fool's pain (of something that feels like pain but isn't, because the underlying physical essence is absent), combined with the conceptual possibility of pain without C-fiber firing (I can conceive of a situation in which I think I am in pain, though my C-fibers are not firing), conceptually rules out pain having C-fiber firing as an underlying physical essence (which it would have if the identity theory were true) (Kripke 1981, lecture 3). Has Kripke here identified a genuine conceptual obstacle to physicalism? Perhaps. Or perhaps not: perhaps it will turn out, upon closer examination, that there is no such obstacle here. The only way to show that, however, will be through logical and conceptual work. Just as in the case of our puzzle about whether only four people might be at the family gathering and the puzzle about jumping into one and the same river twice, a solution will require that we engage, not in empirical investigation, but in reflective armchair inquiry. Establishing more facts about and a greater understanding of what happens in people's brains when they are in various mental states and so on will no doubt be scientifically worthwhile, but it won't, by itself, allow us to answer the question of whether there exists such a *conceptual* obstacle.

So, many philosophical problems—from some of the most trivial to some of the most difficult—appear to be essentially conceptual in nature, requiring armchair, conceptual work to solve. Some are solvable, and indeed have even been solved (the puzzle about the river). Others aren't solved, though perhaps they might be. On the other hand, it might turn out that at least some philosophical problems are insoluble, at least by us. Perhaps, as some "mysterians" propose, while some philosophical problems have solutions, they are solutions that lie beyond our human cognitive and conceptual grasp. Alternatively, perhaps some philosophical problems have no solution, *period*, because the problems result from certain fundamental conceptual commitments that are either directly irreconcilable or generate unavoidable paradoxes when combined with certain empirically discovered facts.

So there are perfectly good questions that demand answers, and that can be answered, though not by empirical means—let alone by that narrower form of empirical investigation referred to as the scientific method. In order to solve many classic philosophical problems, we'll need to retire to our armchairs, not to the lab.

But is that all there is to philosophy? What of the grander metaphysical vision traditionally associated with academic philosophy? What of plumbing the deep, metaphysical structure of reality? That project is often thought to involve discerning, again by armchair methods, not what *is* the case (that's the business of empirical inquiry) but what, metaphysically, *must* be so. But how are philosophers equipped to reveal such hidden metaphysical depths by sitting in their armchairs with their eyes closed and having a good think?

I suspect this is the main reason why there's considerable suspicion of philosophy in certain scientific circles. If we want to find out about *reality*—about how things stand outside our own minds—surely we will need to rely on empirical methods. There is no other sort of window onto reality—no other knowledge-delivery mechanism by which the nature of that reality might be revealed.

This is of course a traditional empiricist worry. Empiricists insist it's by means of our senses (or our senses enhanced by scientific tools and techniques) that the world is ultimately revealed. There is no mysterious *extra* sense, faculty, or form of intuition we might employ, while sat in our armchairs, to reveal further, deep, metaphysical facts about external reality.

If the above thought is correct, and armchair methods are incapable of revealing anything about the nature of reality outside our own minds, then philosophy, conceived as a grand metaphysical exploration on which we can embark while never leaving the comfort of our armchairs, is in truth a grand waste of time.

I'm broadly sympathetic to this skeptical view about the value of armchair methods in revealing reality. Indeed, I suspect it's correct. So I have a fairly modest conception of the capabilities of philosophy. Yes, I believe we can potentially solve philosophical puzzles by armchair methods, and I believe this can be a valuable exercise. However, I am suspicious of the suggestion that we philosophers construe what we then achieve as our having made progress in revealing the fundamental metaphysical nature of reality, a task to which I suspect such reflective, armchair methods are ultimately hopelessly inadequate.

So perhaps there's *at least this much right about scientism*: armchair philosophical reflection *alone* can't reveal anything about reality outside our own minds. However, as I say, that doesn't mean such methods are without value. After all, scientists sometimes employ the same methods, and with scientifically valuable results. Galileo is credited with constructing a thought experiment by which he established that the Aristotelean theory that heavier objects fall faster than lighter ones in direct proportion to their weight is mistaken. Galileo noted that Aristotle's theory predicted that two balls, one heavier than the other, should fall at different speeds: the heavier falling faster. This theory could be empirically tested, of course, and some suppose Galileo tested it by dropping objects from the top of the Leaning Tower of Pisa. However, Galileo himself records no such experiment. What he did do was perform a thought experiment. He imagined the hypothetical balls of different weights now chained together. Being now combined into a single, even heavier object, Aristotle's theory predicts that the balls combined should fall even faster than they did individually. However, because the lighter ball previously fell more slowly than the heavier, Aristotle's theory also predicts that the lighter ball should pull tight on the chain behind the heavier ball and act as a brake on it. But then the balls combined should fall more slowly than did the heavier individually. In short, Aristotle's theory, in combination with some modest assumptions about the effects of the chain, generates a *logical contradiction*. Thus, it cannot be true.

Of course, Galileo's thought experiment did not allow Galileo to predict what would happen when such balls are released. It didn't reveal how reality is: such objects, when released, might fall at the same speed, at different speeds, or transform into a pair of doves and fly off into the sunset—we can't know what they will or won't do without observing at least some instances. It just revealed that Aristotle's theory, insofar as it contains or generates logical contradictions, can't constitute a correct description of it.

So while thought experiments and other forms of armchair reflection can't tell us anything substantive about how things stand in reality, they can at least allow us to make significant scientific progress by showing that certain theories about reality are false. Is Galileo's thought experiment an example of science, or of philosophy? Insofar as it targets a scientific theory—a theory about how physical objects behave—perhaps it belongs more properly to science. However, note that the same armchair method employed by Galileo is also regularly employed by philosophers (for further discussion of the legitimacy of armchair methods, see Sorrell, this volume).

Richard Dawkins on Science and Philosophy

I turn now to Richard Dawkins. To what extent is Dawkins wedded to scientism? In 2013, I engaged in a public discussion with him on this subject (Dawkins and Law 2013). Dawkins began by acknowledging that "philosophy does seem to me to be a subject" (he contrasted philosophy with theology, which he does not consider a subject, and which he has elsewhere likened, much to the annoyance of many theologians, to fairy-ology). Moreover, when presented with an example of a conceptual puzzle similar to my example above involving familial relations, Dawkins said that "that's something you get out a pencil and paper . . . you *do a scientific investigation in your head*" (my italics). According to him, this sort of armchair, conceptual activity *is science*: "Plenty of good science goes on in armchairs." When I suggested to him that the mind-body problem was, at root, a conceptual puzzle that cannot be solved by empirical inquiry, a problem that will be solved, if at all, by armchair methods, Dawkins responded, "That has the ring of sense to me" (at 39 min. 50 sec.). Further, when I suggested that given my understanding of what philosophy should be, "maybe I'm going to turn out to be a scientist," he replied, "Well, of course you are" (at 35 min.).

So Dawkins is not dismissive of philosophy as I understand it and in the form I would wish to defend it. He also acknowledges that, for example, the mind-body problem is a philosophical puzzle—a puzzle we will need to employ armchair, conceptual methods to solve if it's to be solved at all. However, his conception of science is sufficiently flexible to encompass all these activities, so that even philosophers engaged in purely conceptual, armchair inquiry qualify as scientists. Having said that, I suspect that insofar as philosophy is understood as an activity aimed at revealing how things stand in reality, Dawkins would dismiss armchair philosophy as a waste of time. But then I suspect he'd be right to do so.

Not only does Dawkins unambiguously acknowledge that "perhaps there are some genuinely profound and meaningful questions that are forever beyond the reach of science" (2006, 80), but he also seems happy to concede that moral questions may fall into this category. About moral questions, he says, "We can all agree that science's entitlement to advise us on moral values is problematic to say the least" (ibid.) (he may not even consider such questions answerable at all).

We have looked at two kinds of questions—moral and philosophical—that Dawkins appears to be prepared to acknowledge that empirical methods can't, or probably can't, answer, and that (at least in the case of some philosophical questions) *other methods might*. I don't doubt there are other examples of questions that science (understood as a form of empirical inquiry) alone cannot, or cannot fully, answer but other methods might.[1] However, I'll stick with just these putative illustrations.

The Veil

I turn now turn to religious, new age, and popular divine or supernatural belief, and the suggestion that those who are critical of such beliefs from a scientific perspective are often guilty of scientism.

A thought often employed by those who make the accusation of scientism in defense of such beliefs is that reality is divided by something like *a veil*. On one side of the veil lies the empirically observable, investigable, material world. On the other side lies a divine or supernatural realm. This realm is variously supposed to be populated by beings such as the deceased (who are supposed to have "passed over" to the other side not just figuratively but literally), angels, spirit guides, demonic beings, and of course gods. Occult forces and energies are also supposed by some to operate behind such a veil. Knowledge of what "lies beyond" is often supposed to depend on the ability of at least some of us to peer, if only dimly, through this veil and obtain glimpses of what's on the other side. Usually, some sort of supersense is invoked. TV's Psychic Sally, for example, claims that she can sense the presence of those who have "passed over" and can communicate with and relay messages back from the other side. Others believe they are able to sense the presence of guardian angels or other spirit guides or beings. Sometimes, the "doors of perception" to what lies beyond are supposedly opened with chemical assistance. The Delphic oracle of ancient Greece received communications from the god Apollo while perched on a tripod placed over vapors rising from a chasm. Some religious people believe also that at least some of us possess a reliably functioning god-sense or *sensus divinitatis* by means of which the Judeo-Christian God reveals himself. Many self-styled "spiritual" folk claim to have experienced some sort of transcendent, spiritual realm. It's by means of such subjective experiences, and perhaps also signs given and wonders performed on this side of the veil by those on the far side, that we can obtain at least some knowledge of those residing "beyond." The empirical sciences, by contrast, are incapable of penetrating the veil. The proper province of those sciences lies on this side of the veil. There's usually acknowledgment, from those who believe in the veil, that their beliefs concerning what lies beyond cannot be "proved" by scientific means. But then, they quickly add, neither can these beliefs be refuted by such means. The veil effectively immunizes beliefs about what lies beyond against any sort of scientific, empirical refutation.

A version of this thought that there are these two domains, of which the methods employed by empirical science are suited to investigate only one, is presented by the paleontologist Stephen J. Gould, who maintains that science and religion are "non-overlapping magisteria." Science is concerned with the age of rocks, for instance, whereas religion is focused on the "rock of ages." Religion is not well equipped to address questions about, say, how Earth was formed and life appeared. That is the proper business of science. But then, similarly, the empirical sciences are in no position to address questions about meaning and value or the existence of, say, souls or God. Such "big questions" are the proper province of religion, not science. Gould writes that science

> tries to document the factual character of the natural world, and to develop theories that coordinate and explain these facts. Religion, on the other hand, operates in the equally important, but utterly different, realm of human purposes, meanings, and values—subjects that the factual domain of science might illuminate, but can never resolve. (Gould 2002, 4)

He specifically mentions the existence of souls as something the investigation of which lies beyond the proper remit of science:

But I also know that souls represent a subject outside the magisterium of science. My world cannot prove or disprove such a notion, and the concept of souls cannot threaten or impact my domain. (Gould 2012, 575)

On Gould's view, science and religion are not in, and cannot come into, conflict. Science is no threat to religious belief, properly understood, though of course it is a threat to young-Earth creationist beliefs about the age of rocks. Young Earth creationism is an example of religions entering territory that is the proper province of science.

Is it true that beliefs concerning gods, souls, the supernatural, and so on are immune to refutation, or at least to scientific refutation? No. Here are three reasons why.

First, it is possible to refute beliefs about what lies beyond the veil by showing that the beliefs involve some sort of *logical contradiction or conceptual muddle* (a point Gould might of course concede). Suppose someone claims to have discovered a round square in the jungles of Brazil. Will the world's mathematicians mount an expedition to Brazil to investigate? Of course not. They can know, from the comfort of their armchairs, that there's no such thing in the jungles of Brazil. In the same way, we might refute a scientific theory by showing that it generates logical contradictions (we have already noted Galileo's armchair refutation of Aristotle's theory about falling objects). But then god beliefs might also generate contradictions of this sort. Take belief in some sort of divine agency that is supposed to have beliefs and desires on which it acts, but also to be nontemporal: a being that does not exist in time, as we do, but is the creator of time. Perhaps we can show this belief doesn't make sense, given that psychological states such as belief and desire necessarily have temporal duration. In which case, we might refute, from the comfort of our armchairs, belief in such a deity, despite the fact that the deity in question supposedly resides behind the veil (I don't claim that this particular refutation succeeds; I merely point out that such refutations of beliefs regarding what lies behind the veil are at least possible). Would such a refutation be scientific? I'd call it philosophical, but note that we have already seen Dawkins's use of *science* is sufficiently elastic to allow such armchair, conceptual refutations to qualify as scientific.

Second, beliefs about what is behind the veil, though not directly observable, are *potentially empirically refutable*, given they have consequences for what we should or should probably observe on this side of the veil. Scientists regularly confirm and disconfirm hypotheses about unobservables. Hypotheses about the distant past of this planet—such as that dinosaurs once roamed Earth—also concern tracts of reality that are necessarily unobservable by us. Yet such hypotheses can be confirmed, and indeed refuted, by observation. That's because there are things we would much more likely observe (a certain kind of fossilized skeleton, etc.) if the hypothesis were true than if it were false. Similarly, hypotheses about subatomic particles concern tracts of reality that are unobservable, but again we can observationally confirm or refute hypotheses about electrons, the Higgs boson, and so on because these hypotheses also have consequences so far as what is empirically observable.

But then, similarly, hypotheses about what lies behind the veil in some divine or supernatural realm may be empirically confirmed or refuted insofar as they, too, have consequences so far as what we should expect to see on this side of the veil. If I posit a cosmic doodah behind the veil, and say no more about it than that it's an ineffable thingummyjig, it appears that I've made a claim that's hard to refute observationally, given that it has no obvious observational consequences. However, as we add more content to our beliefs about what "lies beyond," those beliefs do become potentially empirically confirmable or refutable. Belief in a god that answers petitionary prayers, for example, has been scientifically investigated. Two large-scale, multi-million-dollar studies of the effects of petitionary prayer on heart patients—one performed under the leadership of Herbert Benson, a cardiologist who previously suggested that "the evidence for the efficacy of intercessory prayer . . . is mounting" (Benson et al. 2006)—found that prayer had no beneficial effect on patients. This was not mere *absence of evidence* of the beneficial effects of prayer but *evidence of the absence* of such effects, and thus evidence against the existence of a god who produces such beneficial effects in response to petitionary prayer.

Despite the fact that they are considered to concern powers, beings, and so on behind the veil, various new age, spiritual, and other claims have also been scientifically investigated. Take, for example, the claim that meditating while holding crystals can, by means of some power operating behind the veil, bring about higher concentration powers and energy levels and increased spiritual well-being. That meditating while holding crystals has such effects was scientifically investigated in a double-blind controlled experiment conducted at the Anomalistic Psychology Research Unit, Goldsmith's College, University of London (French, Williams, and O'Donnell 2001). No such effect was found (no difference in response from those holding real crystals and those holding fake crystals). Again, this was evidence against the existence of any such power operating behind the veil.

Other claims concerning the divine also appear to be evidentially challenged. Take, for example, the claim that this world is the creation of an omnipotent, omniscient, and supremely *malevolent* deity whose malice is without end. If there were such a deity behind the veil, surely things should look very different on this, the empirically observable side? Yes there's pain and suffering, but surely there's far too much love, laughter, ice cream, and rainbows for this to be the creation of such a supremely wicked being. But if we can, quite reasonably, rule out such a deity on the basis of observation, why can't we similarly rule out the existence of an omnipotent, omniscient, and supremely *benevolent* deity on much the same basis? The answer is that there's far too much suffering in this world for it to be the creation of such a god. Assuming this is an effective refutation of these two god beliefs,[2] is it a *scientific* refutation? It's certainly empirically based. However, if we understand a scientific refutation to involve the scientific method and scientific institutions developed over the last few hundred years, then it appears not to be scientific. It's a variety of that still highly effective commonsense kind of refutation that might have been offered long before the development of those methods and institutions. Another example: I might effectively empirically refute your claim to have a cat under your shirt by carefully going around you and patting your shirt. If I find no suspicious lumps and hear no "meows," it's reasonable for me to believe there's no cat there, despite my not having actually looked under your shirt. Is this successful refutation of your claim scientific? I'd say not.

This is not to the say that the argument from evil against the existence of an omnipotent, omniscient, and supremely benevolent god can't take a scientific form. Much of the suffering the world contains has been revealed by science. Examples include hundreds of millions of years of animal suffering before humans have shown up, and two hundred thousand years of approximately a third to a half of every generation of human children dying (typically pretty horribly, from disease, malnutrition, etc.) before reaching the age of five.

Another way in which science, or empirical inquiry, might threaten beliefs concerning what lies behind the veil is by providing evidence that we are highly prone to error when it comes to forming such beliefs on the basis of, say, subjective experiences involving a sense of presence (much like the experiences TV's Psychic Sally reports). Perhaps, if I seem to see a snake lying in front of me, it is, other things being equal, reasonable for me to believe there's a snake there. But if I am then told by a reliable source that I have been given a drug that causes highly convincing-looking snake hallucinations, it's no longer reasonable for me to believe there's a snake before me. Of course, there might still be a snake present, and indeed, my belief that there's a snake might have been brought about by my reliably functioning perceptual faculties (perhaps I have been misled about having been drugged and am just observing a real snake in the usual way), but still, given this new evidence, surely it's no longer reasonable for me to continue to believe there's a snake there. But then, similarly, science might reveal that we humans are constitutionally prone to false beliefs in the presence of invisible agents (gods, ghosts, spirits, angels, demons, the deceased, etc.) when those beliefs are merely grounded in a subjective sense of presence.[3] Under those circumstances, surely I can no longer reasonably maintain my belief in such an agent given only such a subjective sense of presence. Here is evidence, not that there are no gods, ghosts, or whatever, but that my subjective sense that there are can no longer be trusted. If a subjective sense of presence is my only basis for belief, then I should cease believing.[4]

To sum up, while there may be "limits to science," certainly when understood as a form of empirical inquiry grounded in methods and institutions developed only over the last few hundred years or so, it does not follow that religious beliefs, new age beliefs, beliefs about the supernatural, and so on are immune to refutation, let alone scientific refutation.

Dawkins's Argument in *The God Delusion*

Dawkins's book *The God Delusion* has provoked many accusations of scientism, some of which I'll examine shortly. The central aim of the text is to show that there very probably is no God. Dawkins is critical of design arguments—such as fine-tuning arguments—that aim to justify belief in God by pointing to features of the universe that might seem highly improbable if not a product of some sort of intelligent design. In response to such arguments, Dawkins insists that when theists appeal to God to explain such otherwise supposedly improbable features of the universe, they overlook the fact that the god to which they appeal must be at least as complex, and thus at least as improbable, as that which he is invoked to explain:

> A designer god cannot be used to explain organized complexity because any god capable of designing anything would have to be complex enough to demand the same kind of explanation in his own right. God presents an infinite regress from which he cannot help us escape. (Dawkins 2006, 136)

If the existence of the universe having such features by chance is highly improbable, then, says Dawkins, the existence of a god having the kind of complexity to account for it must be even more improbable. So by introducing God, we merely postpone the problem of accounting for such complexity. But then the universe's complexity provides no justification for introducing God. Worse still, if the theist is right and the probability of such complexity just happening to exist is low, then, by the same logic, *the probability of God existing must be even lower.*

Dawkins suggests this argument is scientific, but is it? Why does Dawkins suppose God must be at least as complex as that which he is invoked to explain? Perhaps because, if something is to qualify as having been designed by God, it must be represented in the mind of God, and the structural complexity of what is designed must be there in the representation if it is to qualify as a designed feature. However, this thought about what must be true of representations appears to be delivered by armchair, conceptual reflection on the nature of representation rather than by empirical inquiry. In which case, isn't Dawkins's argument *philosophical rather than scientific*? Perhaps not: because Dawkins classes even such armchair philosophical reflection as science, his argument still qualifies as *scientific*, as Dawkins uses that term.

Is Dawkins's very elastic use of *science*—on which even armchair philosophical reflection qualifies—legitimate? I don't particularly object to his using the term in this way so long as he is clear about what he means. The fact that Dawkins's organization is called the Richard Dawkins Foundation for Reason and Science suggests that even he understands reason to extend beyond the scope of mere science (otherwise, why not just call it the Richard Dawkins Foundation for Science?). So where, according to Dawkins, *does* science end and the rest of what falls under the umbrella of reason begin? That isn't clear to me.

Is Dawkins's argument cogent? I won't assess it here. A number of theists have attempted to refute it, including Richard Swinburne, William Lane Craig, and Alvin Plantinga. My interest lies, not in attempts to directly refute Dawkins's argument, but in the charge of scientism.

McGrath's Response to *The God Delusion*

The theologian Alister McGrath, author of *The Dawkins Delusion?*, is one of the leading religious critics of *The God Delusion*. McGrath has penned numerous articles attacking Dawkins's book and argument. A fairly typical example is "The Questions Science Cannot Answer—the Ideological Fanaticism of Richard Dawkins's Attack on Religious Belief Is Unreasonable to Religion—and Science," published in the London *Times* in 2007. In that article, McGrath begins by pointing out that even scientists such as Peter Medawar acknowledge there are questions science can't answer:

> In *The Limits of Science*, Medawar reflected on how science, despite being "the most successful enterprise human beings have ever engaged upon," had limits to its scope. Science is superb when it comes to showing that the chemical formula for water is H_2O. Or, more significantly, that DNA has a double helix. But what of that greater question: what's life all about? This, and others like it, Medawar insisted, were "questions that science cannot answer, and that no conceivable advance of science would empower it to answer." They could not be dismissed as "nonquestions or pseudoquestions such as only simpletons ask and only charlatans profess to be able to answer." This is not to criticise science, but simply to calibrate its capacities. (McGrath 2007)

McGrath then accuses Dawkins of being ideologically committed to science, of supposing that "science has all the answers." However, we have already seen that Dawkins clearly acknowledges in the book McGrath is here attacking that "perhaps there are some genuinely profound and meaningful questions that are forever beyond the reach of science," and suggests that moral questions may fall into this category. So McGrath is attacking a straw man. We have seen that Dawkins allows room for philosophy, too—though his understanding of science is broad enough to allow even a priori armchair philosophical theorizing and reflection to qualify as science.

McGrath's charge of scientism not only misrepresents Dawkins. It is also, more to the point, *irrelevant to the question of whether or not Dawkins's argument against the existence of McGrath's God is cogent*. Suppose McGrath is right, that there are questions science cannot answer, and that other methods can. Does it follow that Dawkins's argument against the existence of God fails? Clearly not. Dawkins's argument might still be sound.

McGrath then attempts to refute Dawkins's argument against the existence of God, not by identifying some specific flaw in it, but by simply insisting we can't prove there's no God. However, Dawkins himself points out in *The God Delusion* that McGrath's earlier attack on Dawkins seems to turn on the "undeniable but ignominiously weak point that you cannot disprove the existence of God" (2006, 80). In *The God Delusion*, Dawkins clearly and explicitly *agrees* with McGrath that we can't conclusively "prove" there's no God, but then points out that that doesn't mean that belief in God is immune to scientific skepticism. For, as we have already noted, God hypotheses can have observable consequences: "a universe with a superintendent would be a very different kind of universe from one without. Why is that not a scientific matter?" (2006, 80) Dawkins notes in *The God Delusion* that in response to this question, McGrath previously offered no answer. It's ironic, then, that even in this later attack now aimed squarely at *The God Delusion* itself, McGrath still offers no answer.

Elsewhere, McGrath does at least explain why he supposes there can at least be no conclusive "proof" or "disproof" of the existence of God:

> Any given set of observations can be explained by a number of theories. To use the jargon of the philosophy of science: theories are under-determined by the evidence. The question then arises: What criterion can be used to decide between them, especially when they are "empirically equivalent." Simplicity? Beauty? The debate rages, unresolved. And its outcome is entirely to be expected: the great questions remain unanswered. There can be no scientific "proof" of ultimate questions. Either we cannot answer them, or we must answer them on grounds other than the sciences. (McGrath and McGrath 2007, 14)

The suggestion seems to be that, consequently, the "ultimate question" of whether or not God exists cannot be settled by science (by the way, notice McGrath understands science more narrowly than Dawkins—he supposes that what science can settle is settled, in every case, by observational evidence).

But are "God exists" and "God does not exist" empirically equivalent? Do these two worldviews fit the observational data equally well? Dawkins's point is of course that they are *not* empirically equivalent. Given that hypotheses about unobservables (subatomic particles, the distant past of this planet, etc.) can indeed be empirically confirmed and disconfirmed beyond reasonable doubt by reference to what is observable, why shouldn't God beliefs also be similarly empirically confirmable or disconfirmable? Given that we can empirically disconfirm the belief that there's a God who answers petitionary prayers, and that the belief in a God that is omnipotent, omniscient, and omni*malevolent* also appears to be empirically disconfirmable, why should we suppose that McGrath's belief in a God that is omnipotent, omniscient, and omni*benevolent* is not empirically disconfirmable? McGrath offers no explanation. He simply declares, without argument, that his God is off-limits to science.

William Reville on Religion and Scientism

In a 2014 article titled "Philosophers Must Oppose the Arrogance of Scientism," William Reville, a professor emeritus of biochemistry, defends religion against Richard Dawkins and others whom Reville accuses of scientism. He believes that these scientists dismiss religion because they are fundamentalist materialists who think that "nothing exists but the material." However, "materialism is a philosophy that has not—and probably cannot—be proven." Reville adds:

> It is reasonable to be a materialist. But, since materialism is unproven, materialists must accept that, no matter how improbable it seems to them, there is a possibility they might be wrong and a supernatural dimension might exist. Materialists are therefore obliged to respect the position of religious people who believe in the supernatural but accept all that science has and will discover. (Reville 2014)

Reville thinks the supernatural is off-limits to science:

> Science studies the natural world. It is materialistic in its method but not in its philosophy. Science does not deny the supernatural, it simply has nothing to say about it. Science and religion address different aspects of reality and do not contradict each other, as noted by the eminent science writer Stephen J Gould in his book *Rocks of Ages*. (Ibid.)

Now, as we have seen, Dawkins actually justifies his rejection of belief in God by an *argument*. Reville does not address or even mention that argument, but instead simply declares that God, being supernatural, is off-limits to science.

Perhaps Reville is right and it is *possible* a supernatural realm exists. However, that's not to say it's remotely *probable*. Many supernatural claims are highly improbable, and in some cases have been shown to be so through an application of the scientific method. The supernatural is not off-limits to science. But then, if supernatural claims are often empirically investigable and can in fact be empirically refuted, why shouldn't Dawkins and others succeed in refuting—perhaps even in *empirically* refuting—a particular God claim? Reville does not explain.

He does encourage philosophers like me to make it clearer to scientists that science has limits:

> One very important function of philosophy is to identify scientific questions. This is important in order to keep science from going off the rails. Philosophy is not doing its job. (Reville 2014)

While as a philosopher I am happy to acknowledge that science has its limits, I see no reason to agree with Reville that supernatural claims are in principle off-limits to science. Indeed, they are not off-limits, and neither are many God claims.

Some Other Illustrations

I have provided two illustrations of how the charge of scientism has been made in a baseless and indeed irrelevant way against critics of religious and/or supernatural beliefs. It is not difficult to find further examples.

Here, for example, is Bishop James Heiser:

> The efforts of scientists to disprove the existence of God is not a pursuit of *Science*, but *Scientism*. (Heiser 2012)

As should now be clear, efforts to disprove the existence of God do not necessarily involve an embrace of scientism.

In his essay "Has Science Disproved God?," Barry L. Whitney writes:

> It is important to note that science, unlike scientism, should not be a threat to religious belief. Science, to be sure, advocates a "naturalistic" rather than "supernaturalistic" focus, and an empirical method for determining truths about the physical world and the universe. Yet the proper mandate of

science is restricted to the investigation of the natural (physical, empirical dimension) of reality. It is this restriction that scientism has violated. (Whitney 2006, 16)

This is another example of an author just assuming that the supernatural is off-limits to science and then using that assumption to immunize his religious belief against any potential scientific threat. As we have seen, empirical science is more than capable of investigating many supernatural claims, and has done so successfully. To suppose otherwise is not to be guilty of scientism.

The G. K. Chesterton scholar Dale Ahlquist writes:

> Too often a prominent physicist or biologist is believed when he declares that empirical science has disproved the existence of God. . . . The fallacy is, of course, that empirical or experimental science is limited to the work of discovering and applying truths about the material world. If there is a spiritual presence in the material world, physical science will not discover it; and if we discover it, physical science will have no idea of what it means. Chesterton would have none of such scientism.
>
> *What can people mean when they say that science has disturbed their view of sin? Do they think sin is something to eat? When people say that science has shaken their faith in immortality, do they think that immortality is a gas?* (Ahlquist 2014)

But immortality doesn't have to be material—and certainly doesn't have to be a *material substance* like a gas—in order to be scientifically investigable. Neither does God. Again, an unjustified and mistaken characterization of the remit of science is used by the author to immunize religious belief against any potential scientific refutation.

In "Has Science Done Away with God?," the Catholic apologist Matt Fradd asks:

> Hasn't science disproved God? No, and it is not within the ability of science to do so. Science is a method that one can use to discover information about the natural world. . . . Examining God's material creation using a method which, by its very nature, is limited to the material universe cannot provide evidence against the existence of an immaterial God. . . . The view that science can or should provide the answer to every question is known as *scientism*. (Fradd 2012)

Fradd asserts that it is beyond the remit of science to disprove God. But no justification for this claim is given. We're just told those who suppose otherwise are guilty of scientism.

Conclusion

In the hands of some—including many theologians—the charge of "scientism!" has become a lazy, knee-jerk form of dismissal, much like the charge of "communism!" used to be. It constitutes a form of rubbishing, allowing—in the minds of those making the charge—for criticism to be casually brushed aside. No doubt some things really are beyond the ability of science, and perhaps even reason, to decide. But there's plenty that does lie within the remit of the scientific method, including many religious, supernatural, new age, and other claims that are supposedly "off-limits." However, because the mantra "But this is beyond the ability of science to decide" has been repeated so often with respect to that sort of subject matter, it is now heavily woven into our cultural zeitgeist. People just assume it's true for all sorts of claims for which it is not, in fact, true. The mantra has become a convenient, immunizing factoid that can be wheeled out whenever a scientific threat to belief rears its head. When believers are momentarily stung into doubt, many will attempt to lull them back to sleep by repeating the mantra over and over. The faithful murmur back: "Ah, yes, we forgot—this is beyond the ability of science to decide. . . . zzzz."

······

Notes to Chapter Seven

1. So, for example, I am not convinced by those forms of scientism insisting that, say, humanities subjects like history employ nothing but the scientific method understood narrowly, so as to exclude the kind of understanding one acquires through an imaginative identification with others, in some cases achievable only through the use of music, poetry, painting, and other forms of artistic expression. While there's much that science can contribute to history as a discipline, that's not to say that history is, or should be replaced by, science narrowly conceived. For further discussion of this issue, see Kitcher in this volume.

2. Notwithstanding the construction of theodicies and appeals to "skeptical theism"

that might, in each case, be invoked to try to deal with this apparent evidence against these two god claims, I believe it is.

3. That we humans are indeed constitutionally prone to false belief in invisible agency has been noted by many scientists, including the evolutionary psychologist Justin Barrett, who has also postulated a mechanism to account for this tendency—an evolved Hyper-Active Agency Detecting Device. See Barrett 2004 and Boyer 2002.

4. Notice my argument here is not that where a religious or other seemingly supernatural experience can be correctly *explained naturalistically* (in the style of Barrett and Boyer, for example—see preceding note), it ought not to be trusted. Such naturalistic debunking arguments may or may not be cogent. Rather, my argument is a *defeater argument*: if I have good grounds for thinking that in the circumstances in which it appears to me that P, that appearance is quite likely to be misleading, then I ought not to rely on that appearance—I ought not to believe that P. Such defeater arguments don't require my particular experience to *actually be produced* by an unreliable mechanism (I might *actually* be seeing a real snake via a reliable perceptual mechanism, and not be drugged), or indeed have a correct naturalistic explanation (God might, on this occasion, *actually* be supernaturally revealing himself to me). Even if I am actually perceiving a snake, or am actually experiencing God, by what is in actual fact a reliable belief-grounding mechanism, the fact remains that I *shouldn't* believe given the evidence that I'm in a situation where such experiences are very often deceptive.

Literature Cited

Ahlquist, D. 2014. "Scientism." American Chesterton Society, http://www.chesterton.org /scientism/.

Barrett. J. L. 2004. *Why Would Anyone Believe in God?* Lanham, Maryland: AltaMira Press.

Benson, H., et al. 2006. "Study of the Therapeutic Effects of Intercessory Prayer (STEP) in Cardiac Bypass Patients: A Multicenter Randomized Trial of Uncertainty and Certainty of Receiving Intercessionary Prayer." *American Heart Journal* 151: 934–42.

Boyer, P. 2002. *Religion Explained: The Evolutionary Origins of Religious Thought.* New ed. London: Vintage.

Dawkins, R. 2006. *The God Delusion.* London: Bantam.

Dawkins, R., and S. Law. 2013. "In Conversation with Richard Dawkins—hosted by Stephen Law." YouTube video, posted by ThinkWeekOxford March 13, 2013, 1 hr. 29 min. https://www.youtube.com/watch?v=zvkb iElAOqU.

Fradd, M. 2012. "Has Science Done Away with God?" Mattfradd.com (blog), August 7, http://mattfradd.com/has-science-done-away-with-god/.

French, C., L. Williams, and R. O'Donnell. 2001. Unpublished paper presented at the British Psychological Society Centenary Annual Conference, Glasgow.

Gould, S. J. 2002. *Rocks of Ages.* London: Vintage.

———. 2012. "Nonoverlapping Magisteria." In *Philosophy of Religion: An Anthology,* edited by L. Pojman and M. Rae, 568–77. Boston: Wadsworth.

Harris, S. 2012. *The Moral Landscape: How Science Can Determine Human Values*. London: Black Swan.

Heiser, J. 2012. "Will Science Disprove God's Existence?" *New American*, September 21, http://www.thenewamerican.com/culture/faith-and-morals/item/12926-will-science-disprove-gods-existence.

Kripke, S. 1981. *Naming and Necessity*. New ed. Oxford: Wiley Blackwell.

McGrath, A. 2007. "The Questions Science Cannot Answer—the Ideological Fanaticism of Richard Dawkins's Attack on Religious Belief Is Unreasonable to Religion—and Science." *Times* (London), February 10.

McGrath, A., and J. C. McGrath. 2007. *The Dawkins Delusion? Atheist Fundamentalism and the Denial of the Divine*. London: SPCK.

Reville, W. 2014. "Philosophers Must Oppose the Arrogance of Scientism." *Irish Times* (Dublin), January 16.

Whitney, B. L. 2006. "Has Science Disproved God?" In *The Big Argument: Does God Exist?*, edited by J. Ashton and M. Westacott, 15–26. Green Forest, AR: New Leaf Publishing Group.

8

········

Strong Realism as Scientism:
Are We at the End of History?

THOMAS NICKLES

Introduction: Crowbars and Limits

The scientism question is basically whether science overreaches its legitimate
boundaries—and, if so, where and how it does so. This "hubris" question presupposes that there *are* legitimate boundaries, whence the controversy over
how to establish them. I shall make some remarks about limits and then look
at the hubris question in a way that may surprise some readers; for in this
case, the scientists and science studies analysts would appear to be remaining squarely within the proper domain of scientific investigation. Since most
people assume that it is the job of science to tell us the final truth about the
universe, as if it were a direct replacement for traditional metaphysics, the
question of my title sounds odd. They will also be unhappy with my answer to
the question, namely that strong scientific realism, as a general *interpretation*
of the process and products of science, can be termed a form of scientism—
not as dangerous as some forms, perhaps, but scientism nonetheless.

Is science unlimited? No, of course not. Scientists are not omniscient
within their various domains of research; otherwise, no ongoing research
would be necessary. The reader may consult such sources as the July 1, 2005,
special issue of the journal *Science* in order to get the idea. The issue lists a

sample of 125 unanswered questions that puzzle researchers. The scientific enterprise is a mostly progressive *historical* phenomenon that is therefore limited for any immediate future by what has been achieved so far. This is a historically contingent limitation, a de facto, not an ahistorical, a priori, de jure one, since good science, as a project of inquiry rather than a pile of established facts, keeps pushing against its current boundaries. However, there will always be a de facto limitation based on the then-current state of the art, since not even the best scientists can escape their position in history, and each new breakthrough raises additional questions to explore.

Most discussions of the limits of science address the de jure question, the question of "principled" limits, as when authors claim that science is essentially a descriptive-explanatory enterprise and hence can never bridge the is/ought (descriptive/normative) distinction. By contrast, my focus is on practical, historical limits, over the long term. De jure distinctions are usually based, at bottom, on a priori intuitions about what is logically or empirical-lawfully possible, but such a priori exclusions have a poor historical record. There are many instances in the history of science in which what was once considered even *logically* impossible (self-contradictory) turned out not to be, and vice versa (see below). We do not possess Cartesian, godlike powers of reason enabling us to stand outside our historico-cultural standpoint to make such judgments reliably. In this sense, even our judgments of logical impossibility are historically contingent to some degree, historically de facto.

Claiming to know far more than what the latest results support is hubris. While the sciences have taught us a great deal about the place of *Homo sapiens* in the universe, I personally bridle at today's silly claims that the human genome project or brain research finally shows "what it means to be human." What we are learning in these areas is already impressive enough. We don't need the hype. Only additional research will determine how much science has to contribute; and there, I agree, it would be wrong, even cowardly, to stop investigation prematurely on the basis of some traditional "criterion" or religious or political dogma.

Precisely because science is historically changing, constantly expanding, we should not impose limits a priori. So I am not ready to agree with the prominent physicist Lee Smolin (2006) and an older philosophical tradition that work on string theory is unscientific because it is unfalsifiable at present. After all, Wolfgang Pauli's postulation of the neutrino in 1930 was thought by some to be unfalsifiable, but we now have considerable empirical access to neutrinos and even use them as our medium for observing the interior of the Sun (also once thought to be unobservable). But Smolin does make some telling critical points in his effort to explain why physics has not made dramatic progress in a generation. An unjustifiably strong, near-dogmatic commitment to favored research programs helps to explain the widely noted failure of today's science to plunge boldly into terra incognita, where failure is more likely than success—a point to which I shall return.

There may be limitations of a gradual sort that lead to a slowing of fruitful research, even though the science does not hit an immovable wall. Already in the 1960s, the historian of science and technology Derek Price ([1963] 1968) pointed out that since the seventeenth century, there had been an exponential increase in both human power and financial resources poured into enterprises that we now call sciences—and that such a rate of increase could not long continue. More recently, Nicholas Rescher (1978; 2006) developed Price's themes that (1) an exponential increase in such resources is necessary to maintain a linear increase in first-rate discoveries, and hence that (2) the flattening of the exponential curve will result in a decline in the rate of transformative discovery, which may be happening already.[1] For example, how much longer can high-energy physics remain a robust empirical science, given the exponentially increasing financial costs of large accelerators? The US Congress ditched the Superconducting Supercollider project in Texas some years ago despite the fact that billions had already been invested in it. And it now requires a multinational effort to provide resources for the Large Hadron Collider at CERN in Switzerland.

Contingent limits are imposed by scientific conclusions themselves. Our scientific instruments at any given time have limited resolution (Giere 2006), so, although instrumentation will continue to improve, scientists will never justifiably eliminate error bars from graphs of the empirical data. Some contingent, empirical scientific claims considerably trim traditional scientific goals of achieving deep truth and the explanatory understanding that supposedly follows. Supposing that they do hold up, twentieth-century developments in physics impose limits on what we can expect to accomplish. Relativity theory tells us that the vast majority of the universe is beyond observation. In quantum theory, the Heisenberg Uncertainty Relations come readily to mind, but more important to my purpose is the general lack of deep intelligibility (to us) of what is going on at the quantum level. This is a major reason why the leading philosopher of physics Bas van Fraassen (1980; 2001), for example, shuns a realist conception of science. In this he joins such scientific luminaries as

Niels Bohr and Richard Feynman, key makers of our physics. Mature science itself imposes limits on our scientific aspirations.

The history of the quantum revolution thus points to a double limit. While engagement with the world has its intrinsic limits, set by the world itself (as best we can tell), another limit is that of human cognitive resources. With quantum mechanics we appear to have reached such a limit. Great predictive power does not automatically give us great descriptive and explanatory power, that is, genuine understanding of the actual processes involved, which, after all, is what realism is all about. It is possible that a good deal of the universe will be beyond our comprehension to some considerable degree. By that I do not mean that research should or will stop, only that it will become even more model- and simulation- and prediction-driven and less driven by deep descriptive and explanatory understanding. For what is "really" going on may be beyond human intelligibility.

Speaking of models, philosophers of science and other science studies professionals now appreciate that the theory-centered picture of mechanics rarely applies to sciences other than physics; indeed, it applies fairly well only to parts of physics. Today, talk of models is all the rage. What we might term "the models revolution" in science and technology studies has taught us that all sciences employ models of various kinds, whereas only rarely do we encounter a big, universal theory of the sort familiar from mechanics. For our purposes, the important thing about models is that in nearly every case, they are known from the start to be strictly false via the use of simplification, approximation, abstraction, or idealization (Shapere 1984). Scientists need to make these compromises in order to get fruitful results. Simplifying or approximating highly nonlinear equations may be necessary to make useful calculations at all.

Therefore, the pervasive practice of modeling already contradicts the old, five-step conception of "the scientific method" that we all learned in school: (1) note a puzzling phenomenon, (2) formulate a hypothesis to explain it, (3) derive predictions from the hypothesis, (4) test the hypothesis by performing experiments, (5) regard the hypothesis as confirmed or disconfirmed, according to whether or not the predictions are correct. We were then cautioned that confirmation does not guarantee that the hypothesis is true, but that sufficient confirmation might raise it to the status of a theory. Since models are known to be strictly false from the get-go, what we learned in school about how science is done is badly mistaken. To be sure, many authors still regard models as steps on the way to true theories (e.g. McMullin 1985; Wimsatt

2007, chap. 6), but in some areas it appears unlikely that we can get beyond idealized or only approximate models.

I won't take up that issue here beyond emphasizing that although the models revolution makes it more difficult to talk about *truth* in science, it does not undermine the *reliability* of good scientific research. (Here and later I am using *reliable* and its cognates in an everyday sense, not in the truth-conducive sense of reliabilist epistemologies.) Contrary to what some climate change naysayers seem to think, even our very best science employs models, usually models of many different kinds. Models are among our most valuable, most successful research tools. On the pragmatic pluralist position that I defend, which model we use in analyzing a complex phenomenon or target system should depend on what we want to know about it; and there is no need for the models that we use to capture different aspects of the system for different purposes to be mutually consistent. They don't all have to come together into a single, neat theory. In fact, we sometimes need to use incompatible models even for a single purpose (Winsberg 2010, 73).

Although the use of models has greatly extended what we can know, our need to use models also imposes some limits on what we can claim to know with assurance. As noted above, some key scientific results place limits on what we can know (and what there is to know), and human limits also impose constraints. To make my position vivid, I shall compare research tools with crowbars or pry bars. (I have used this ridiculously simple "crowbar model of method" in public lectures and my classes for many years to illustrate the compromise that it represents.) To be useful, a crowbar must sufficiently engage the world at one end and the human hand at the other. How much engagement is sufficient will of course depend on the task at hand. If the world were tomato soup, the "world" end of the crowbar would get no purchase; and if we were fish, we would have no way in which to engage the "hand" end. As I see it, there is typically a compromise of sorts. Let's call it *the crowbar compromise*. The point here is that the tools that we use to probe the universe will all have limited resolution at the "world" end, but they also must produce results that we can grasp at the human end. These results must be at least moderately intelligible to us, insofar as they are useful. To employ broadly Aristotelian terminology, one "end" of a tool must engage the order of being (external reality), while the other must engage us, confined as we are to the order of knowing. In esoteric contexts of research and creative invention, these orders are typically very different. In the deeper scientific cases, it may be that never the twain shall meet.

The result is a compromise perspective that holds both bad news and good news for scientific realists. The bad news is that they are mistaken in saying that science should aim to achieve the final, representational truth about reality and that we know that it sometimes succeeds in that aim, or nearly enough. I shall argue that strong realism is untenable, not a rational goal (cf. Laudan 1984). But the good news for realists is that the fact that our tools work at all tells us something important about reality. Working well does not usually tell us exactly what the world is like "all the way down," but it can provide strong evidence for what the world is *not* like. As Karl Popper (e.g. 1963) always emphasized, these "failures" actually represent important advances in our knowledge. This, however, remains a modest realism. It is not the "good old" strong and deep realism of scientific and philosophical tradition.

Strong Epistemic Realism, Deep Realism, Textbook Realism

As I shall use these terms, *strong realism* consists of two related components, an epistemological component that I shall call *epistemic realism*, claiming that such-and-such theory, model, explanation, and so on is known to be true, or approximately or most probably true; and an ontological component that I shall call *ontic realism* or *deep realism*. Deep realism claims to have discovered "the secrets of nature," that is, to provide a rich, *intelligible* description and explanation about what is really going on in the universe. They are no longer secrets hidden behind high-level abstract equations about whose precise interpretation the leading experts disagree. Meanwhile, *textbook realism* is the epistemic realism (and sometimes fairly deep realism) found in standard scientific textbooks, a realism that overlays the presentation of past scientific successes of the field in question as simply established facts, an uncritical realism that naïve students are invited to embrace. These realisms are mutually compatible in some of their forms, but not all forms of epistemic realism are compatible with deep realism, as the example of *structural realism* demonstrates.

Structural realists are (or can be) epistemic realists who reject deep realism as quasi-metaphysical and are content to say that it is the formal equations that are (approximately) true. Augustin Fresnel's equations of the 1820s concerning the behavior of light have held up despite the fact that we no longer believe that light consists of waves in an ether medium (Worrall 1989). For structuralists, mathematical invariance under deep scientific change is a mark of (approximate) truth. It is the mathematical structure that counts—Fresnel's theory is Fresnel's equations, not their interpretation in terms of richly postulated entities and processes. The only sort of explanation left for structuralists is the shallow one of computing empirically observable patterns from the minimally interpreted equations. The deep descriptive and explanatory understanding of traditional scientific realism is unavailable to us. The great French philosopher-scientist Henri Poincaré (1952) is the precursor to today's structural realism.

Be it noted that I have little problem with what is sometimes called *intentional realism*—the striving for true descriptions of reality as a personal goal. Although I think this is not the right way to express the goals of science, intentional realism does not entail strong realism. Popper was an intentional realist but—in denying that we can ever know the truth or even nonzero probabilities of general, theoretical statements—definitely not a strong realist in my sense.

All empirical claims are fallible, but it is not simple fallibilism that is my target here. It is claims that go far beyond normal empirical support. Claiming to have the absolute, theoretical truth about a general phenomenon is one example. Claiming to have a general scientific method for generating excellent theories or models in all fields is another. A wildly enthusiastic reductionist claim is yet another. Strong realism about the very large, the very small, or the very complex and nonlinear goes beyond what the various sciences are equipped to tell us—so far, anyway. As noted above, some deep realist pronouncements sound suspiciously like traditional metaphysics. While modern science has largely replaced metaphysics as providing our best account of reality, some traditional metaphysical aspirations are best dropped.

Be that as it may, I want to avoid the position of the noted nineteenth-century German physiologist Emil du Bois-Reymond when he said about certain key questions that he proceeded to list, *Ignoramus et ignoramibus*—"We are ignorant and will always be ignorant." Perhaps so, but we cannot specify these questions in advance with any assurance. In fact, we think that science has already shown du Bois-Reymond to be wrong about some of his questions.

Notice that strong realism, surprisingly, implies claims about the future similar to those of du Bois-Reymond, only on the *opposite* side of the realism/knowledge issue. For strong realists are committed to saying that our present, specific understandings in "mature" scientific domains are now so near the truth that they will still (aside from minor adjustments) be orthodox science five hundred years from now, and forevermore—that our successors will never hit upon a successful replacement for today's celebrated conclusions. Strong realists will claim that this is not ignorance, for viable alternatives do not exist today; but how can they know that they never will or that they do not exist abstractly? *That* is the question. Again, as everyone knows, the past history of science and technology is full of accomplishments contrary to entrenched scientific orthodoxy of the time, breakthroughs once claimed to be physically impossible or even logically impossible. Think of the conceptual breakthrough needed in Newton's time to speak of a "state of motion" (not just of rest, given that *state* implies *static*), of "the origin of species" in Darwin (given that *species* meant a fixed type or kind), or of the unconscious mind from Freud to today's cognitive science, for which nearly all cognition is subconscious. The reverse is also true. Modern science declares impossible many things once thought to be possible.

Strong realisms are at best quasi-scientific positions, frequently espoused by scientists themselves in reporting their findings as well as by science analysts, claims made in the name of science about science itself. So for many people, strong realist claims have the aura of genuine, first-order scientific claims. Do these authors realize how strong these claims are, given that they are, implicitly, making strong predictions about the distant future? Wait! These are not really scientific *predictions* but, rather, *forecasts* or *prognostications*. As the makers of these claims, the burden is surely on the realists to support them with empirical evidence. They do provide some evidence. Tomorrow they may even have evidence from today's future, but "the future" is a long time!

I shall articulate these points in the next section, but first notice that there are also present-day consequences—negative consequences for scientific and technological innovation. For when deployed with dogmatic assurance by those currently in power, strong realism can be a research stopper, something that administrators and grants committees can use to discourage deviant, because iconoclastic, research programs.

Dreams of a Final Theory or Illusions of Realism?

Here are a few of my reasons for rejecting both epistemic realism and deep realism. These are social-psychological reasons that supplement the usual historical, logical, and semantic arguments that nonrealists lodge against the defenders of strong realism. I have already noted, via the crowbar compromise, that we cannot simply assume that we humans are capable of understanding, literally and accurately, all domains of nature "all the way down." We already have good empirical evidence that challenges that assumption. I now describe several cognitive illusions that can make scientific realism seem more attractive than it "really" is. I think there are more than a dozen such cognitive illusions. Here I have space to describe a few of the principal ones.[2] You can interpret the overlap among them as just different ways of saying the same thing; but, psychologically, these different ways become mutually reinforcing.

1. *The overconfidence illusion.* Psychological research shows that most people are overconfident of their knowledge, ability, and degree of control. They underestimate risk and, especially, uncertainty. The Nobel laureate Daniel Kahneman (2011, chap. 20) calls it *the illusion of validity*, one component of which is the WYSIATI assumption: what you see is all there is. It would be naïve to suppose that scientists and science analysts usually escape this cognitive illusion. Even the brightest people underestimate the difficulty of completing major projects. Many announced breakthroughs end up as only modest advances; so, as with reference letters, it is wise to discount such announcements by at least 30 percent. The joke about nuclear fusion research also applies here: achieving net power output from a fusion reactor is always forty years away.

2. *The flat-future or end-of-history illusion.* The prominent social psychologist Daniel Gilbert and colleagues have described what they term "The End of History Illusion":

> We measured the personalities, values, and preferences of more than 19,000 people who ranged in age from 18 to 68 and asked them to report how much they had changed in the past decade and/or to predict how much they would change in the next decade. Young people, middle-aged people, and older people all believed they had changed a lot in the past but would change relatively little in the future. People, it seems, regard the present as a watershed moment at which they have finally become the person they will be for the rest of their lives. This "end of history illusion" had practical consequences, leading people to overpay for future opportunities to indulge their current preferences. (Quoidbach et al. 2013, 96)

As Gilbert remarked to a *New York Times* reporter:

Middle-aged people—like me—often look back on our teenage selves with some mixture of amusement and chagrin. What we never seem to realize is that our future selves will look back and think the very same thing about us. At every age we think we're having the last laugh, and at every age we're wrong. (Quoted in Tierney 2013)

Lead author Jordi Quoidbach added, "Believing that we just reached the peak of our personal evolution makes us feel good" (ibid.).

It seems to me that something like this illusion invites scientists and science observers to be strong realists. Just as our personal futures look uneventful or flat compared with our pasts, so does the future history of science. The future is not available to us, of course; and only with the greatest difficulty, via science fiction, scenario planning, and such, can we concretely imagine future knowledge transformations. Even if we succeed, such a scenario immediately strikes us as implausible because it violates current "knowledge." Worse, most future possibilities lie beyond our present horizon of imagination, beyond present conceivability. Yet the past of our most developed, most *mature* sciences has both disrupted orthodoxy and been beyond the imagination of its sufficiently distant predecessors, so what is the warrant for supposing that the future will be only a modest extension and application of present understandings? That view implies that said sciences are near the end of their intellectual odyssey, near the end of their history.

It is important to note that critics of strong realism are not committed to holding that future scientific revolutions are in the offing (although that possibility is certainly in play, given the dynamical couplings and past history of all our sciences). For slow evolution over a long enough time period of several scientific generations can be as transformative as you please. The fact that realists can establish a kind of continuity between specific theory n and theory $n + 1$ in response to Kuhnian claims of revolution therefore falls far short of establishing long-term persistence of the original theoretical commitments. And to the objection that future research can always continue to delve more deeply into what is now understood, there are two replies. One is the problem of practical resources cited at the beginning of this chapter. The other, more interesting one is that historically, nearly every successful, deeper inquiry has eventually disrupted received beliefs and practices. So, supposing that it continues that long, why should we expect the science of the year 3000, or that of the year 10000, to closely resemble the most mature science of today? To say that in physics today we are only a few steps from the final theory or from knowing the mind of God strikes me as worse than literary hype or historical insensitivity. It is arrogance.[3]

3. *The historical maximum illusion.* This is a specific case of confusing a local maximum with the global maximum. Mountain climbers in a dense fog can tell when they have reached a local maximum only by noting that each direction heads downhill from where they are standing. Research scientists at far frontiers are in even more of a fog since they cannot tell whether they are at a maximum at all, local or not, since tomorrow may bring another step upward. Let's say, then, that they are at a *historical maximum*, higher than all previous history of work in that domain. They can at least tell when they have reached a historical maximum (supposing that they are fully up to date and that their research holds up), but they are far from knowing that it will turn out to be the highest peak.

   Strong realists are "Ptolemaic" (pre-Copernican) in privileging our own historical perspective as if it were absolute, thereby enabling us to determine, in favorable cases, that we are at the global maximum of correct understanding, or only a step or two away. But since we cannot compare present work with either absolute metaphysical reality or with future scientific work, such a determination is impossible. To claim that it is (nearly) global is to doubt that the science in question will be seriously progressive in the future. And to say that about the future implies, in turn, that today's frontiers are no longer deep and challenging.

4. *The maturity illusion.* This is the conflation of past-oriented, relative maturity judgments with absolute maturity judgments. Why does maturity enter the picture? Because realists readily agree that at several points in the past, scientists have thought they had the truth; yet it later turned out that they were badly wrong. What, then, makes today's realism more defensible? Because, realists say, the sciences they champion (usually physics) are now *mature*, whereas they were not before.

   The flaw in this move is easy to see. As in the historical maximum illusion, we usually have good reason for thinking that a present-day science is *more* mature than its predecessors, but this same judgment will be possible by future scientists about *their* science, at which point ours will be dubbed relatively immature. "More mature" does not mean "absolutely mature," as if some special threshold has been passed marking a departure from the mundane world of historical culture.

Claims about scientific maturity are really only historically relative and should be historically indexed. Is there just one point in history at which "modern" physics really took off?

Some realists take the maturity claim to imply that scientific frontiers have now been tamed, given the increased sophistication of our methodologies and instrumentation. This view displays a naïveté about scientific research, supposing that it is virtually all normal research under a paradigm in Thomas Kuhn's sense ([1962] 1970). In fact, whenever scientists attempt to move into a quite novel domain, there is no guarantee that the old methods and practices will work. Such exploratory research is replete with trial and error. And to suppose that current paradigms will last forever begs the entire question about the tenability of realism. Those who claim that there is no more scientific frontier, that it has been vanquished, are guilty of a version of the maturity illusion that I call *the frontier thesis illusion*. For these authors are claiming that the exploration of the frontier is basically over, just as Frederick Jackson Turner did in his famous (and better-warranted) claim about the role of the frontier in American history (1920, chap. 1).

5. *The puzzle illusion.* This is the view that most scientific research is problem solving; that most problems are well structured, as found in textbooks (with correct answers to the even-numbered ones at the back); and perhaps also that scientific problems can be compared to crossword puzzles in the sense that they are so heavily constrained that there is a right answer and only one right answer. The reader takes for granted that the problem is legitimate, correctly formulated, and has a correct answer. Thus, neatly solving a problem encourages the belief that we must have hit upon the truth. (This illusion overlaps the last in that all problems are thought to be like the normal scientific problems that Kuhn labeled "puzzles.") Working scientists as well as science students and philosophers are invited to make this mistake insofar as they look backward, Whiggishly, after the problems have been solved, instead of forward to unsolved problems and open research questions at the frontiers of research where the transformative growth of knowledge occurs.

6. *The Whig fallacy, past and future.* In its familiar form, the Whig fallacy interprets and evaluates past practices and products in the light of present ones. It is a kind of intellectual and cultural imperialism in imposing present understandings, goals, and standards on the past instead of

evaluating past work on its own terms. But a similar mistake (I claim) is made by strong realists in imposing their own understandings, goals, and standards on the *future*. They do this gratuitously and then congratulate themselves for finally having achieved a fully mature science that need not undergo much further change. How else can we conclude that the future will be flat, despite continued research? The end-of-history fallacy combines with the Whig fallacy of interpreting the past in the light of present understandings. Each fallacy serves to conceal the other via this circular reasoning.

What about the weaker realists called *structural realists*? My answer is that structural realists, too, are Whiggish. They tend to judge what was correct and necessary (rather than superfluous) in past science by supposing that our current science is (nearly) correct, and then calling our attention to how past work anticipated the equations in today's textbook. This practice is unhistorical, and it begs the realist question concerning the status of our presently best products of mature science. Fresnel's equations may have survived the transition to Maxwellian electromagnetic theory and beyond, but how do we know they will continue to hold up so well as science advances?

Both moves—evaluation of the past in terms of the present and projection of our present understandings and intuitions onto the future—simply presuppose that our own science is basically correct. So, again, our present perspective is taken to be the fixed point in terms of which all other work is to be evaluated, even though the assessment of the not-yet-extant future is made, well, remotely. And notice that the idea of absolute maturity imposes our claims on the future even when we are careful not to be Whiggish about the past.

7. *The fish illusion.* We are like fish in water in that key parts of our intellectual environment are invisible to us. In some cases, they are not yet articulable by us. Other instances are so obviously true as to seem a priori, so that anyone who questions them is considered weird and then marginalized. Wasn't the old assumption that you can't have waves without a medium rightly taken for granted at the time? Wasn't it also as obvious as anything can be that time and space are distinct, and that traveling forward at exactly 10 miles per hour in a train going exactly 120 miles per hour means that you are moving forward at *exactly* 130 miles per hour? Our deepest intuitions are largely products of our culture and our experience of the everyday world.

Some Objections Answered Too Briefly

Opponents of scientific realism are often thought to be intellectually "soft," having fallen into radical (as opposed to moderate) postmodern social constructivism and relativism. No doubt some are. But it seems to me that the situation is largely just the reverse. Some strong realists talk tough, but they are the ones making the extremely strong, empirically and historically contingent claims about current science—hence the more vulnerable claims. And, in my view, it is strong realists who cannot fully resist the temptations of the above and other cognitive-historical illusions and emotional psychological factors such as the satisfaction of predictive success—factors that go beyond the hard-nosed respect for empirical information and clear logical and mathematical thinking that the sciences, above all, are supposed to honor. We would all like to think that we had glimpsed the Forms before we die, or at least that we had lived at a turning point in history.

Some analysts take unexpected theory reduction (such as Maxwell's unification of electricity, magnetism, and light), successful novel prediction, or even simple predictive precision as strong evidence of truth; but these achievements are also indicators of future vulnerability. They are double-edged swords. A precise prediction raises the standard of precision, increasing the possibility that previously inaccessible empirical discrepancies will come to light in future research. Tight logico-mathematical theory unification removes the buffering, the wiggle room, that was there before, thus increasing the vulnerability to cascading failure; for a difficulty in one place can now easily propagate to many places, signaling the need for a major transformation. (Think of the role of the equipartition theorem in physics in the decades around 1900.) And past theories that we now reject also made novel predictions. To my mind, novel prediction serves better as an indicator of present fertility rather than of truth (Nickles 2006).

I shall answer other objections in Nickles (2017 and forthcoming), but two cry out here for a brief response.

Doesn't your position dampen curiosity about our universe and thus diminish motivation for research into the unknown?

Reply. No, for we are still attempting to find the best answers that we can to questions that we find important; and we will continue to do so, as long as thinking is open and we are curious. Progressives are always trying to improve on current capabilities. Scientific curiosity may sometimes seem shallower than on a deep-realist view, but after all, this has not dulled the incentive to develop a fully quantum theoretic conception of the world.

Doesn't denying that we can know scientific truth play into the hands of the antiscience crowd, the purveyors of scientific uncertainty?

Reply. Not as I see it. If anything, it is more empirically objective and hence more honest to speak in down-to-earth, pragmatic terms of what works best (as far as we know today) than to couch accomplishments in terms of absolute, abstract, ahistorical truth. Besides, the current realist orthodoxy does nothing to stop the naysayers. You may think of my position as treating the products of science as closer to human artifacts such as computers, medicines, accounting methods, and message delivery services. I very much appreciate appeals for respect for truth against the willful ignorance of politically powerful groups (McIntyre 2015). My claim is that when it comes to esoteric scientific matters (among others), this really means respect for relative *truthfulness*, that is, for honesty and trustworthiness, for empirically warranted conclusions of the form "*X* works better than *Y* for purpose *P*." Today's scientific results are far more trustworthy than those of the naysayers. I am definitely not one of the "merchants of doubt" in the sense of Naomi Oreskes and Erik Conway (2010).

Conclusion: Strong Realism as Scientism—and What Is at Stake

We are all familiar with today's problem of scientific hype, especially in reports of world-changing breakthroughs, in both the popular media and top scientific journals. I have argued that strong realism is a more innocent sort of hype about the epistemological status of scientific claims, innocent because there are understandable psychological reasons (in addition to the inconclusive evidential and epistemological arguments that realists offer) for adopting their positions. But hype it is. Historically contextualizing the future, as best we can, helps to expose the cognitive illusions that I have discussed (plus others omitted here), and suggests that we adopt an epistemologically more modest position concerning both the products and the processes of the various sciences—yet a position that clearly discriminates the quality of these processes and products from those touted by the antiscience crowd and by the more radical cultural relativists. Innocent though it may be, strong realism produces scientistic claims that go beyond what evidence and argument entitle us to believe.

I have sketched reasons for preferring a forward-looking, pragmatic conception of science based on reliability and fertility assessment rather than truth. At the frontier, nothing beats the importance of fertility assessment, whether the idea is considered true or false. As for reliability, wouldn't it make more sense to judge a scientific claim against other claims (including other scientific claims) on the basis of degree of reliability or trustworthiness than on something as inaccessible to us as final truth?

Sciences are complex, adaptive innovation systems that attempt to replace current products and processes with better ones, much as technological and other innovative business enterprises do. People make decisions about which products and services are better than others, based on what they can actually do. I think we should make decisions about the products and services provided by the sciences and their associated technologies in a similar way. We don't need to say that the iPhone is true, while the flip phone was not or that one type of communication, say, texting, is true, whereas another, say, e-mail, is false. We don't need to say that automatic gene sequencing is true, while doing it manually is false. It is a question of what works best for the purpose at hand at this point in history. Indeed, for many purposes, classical models still work best. Since the general public makes no clear distinction between science and technology anyway, they are already halfway to this conception. I am with those who regard the sciences as ongoing research processes and products rather than collections of established truths or near-truths.

The position I am defending, therefore, holds that when we are being careful, our claims to scientific success need to be historically indexed. My bottom line is that truth-talk adds nothing to what the sciences and technologies have accomplished, except in an everyday, nonrigorous, "folk" sense of truth, and it provides no additional purchase to frontier research. This is an empirical claim, based on what we have learned about research processes so far. Talk of truth adds no new tools to the researcher's toolkit. It adds nothing to our understanding of scientific practice. So I agree with those who contend that talk of truth in these contexts is really a *façon de parler* for talk about historically indexed, relative assessments of reliability or fertility.

We should not be surprised that old models *of* science need revising, just as old models *in* science often do. Look how long it has taken science studies analysts, as well as many scientists themselves, to correct old, mistaken views about science. The job is not finished. Modern science has been one of the most progressive projects in human history, but that does not mean that we yet understand fully how it works. We humans did it, but we still struggle to understand *how* we did it, and how we do it today.

So what is at stake? My answer: the rate of scientific progress, science education, public education about how science works, and public policy.

The strong realism that I have criticized in this chapter distorts our conception of science in ways that affect science itself as well as its reception by laypersons, including legislators and administrators. To say that the current results of our mature sciences are true, or nearly so, implies that they are near the end of their history, thus increasingly sterile. Such a view induces a conservatism that can extend to science policy and funding, by making it even more difficult for proposals that violate current orthodoxy. The sciences have come under heavy criticism in recent years by many scientists, as well as by science analysts, for institutionally discouraging transformative change, for shunning daring projects that, given the uncertainties involved, will probably fail but that have a chance to make breakthroughs. No doubt those politicians who have the power of the purse are part of the problem. (Failing to understand the uncertainties of frontier research, they often use the "uncertain science" tactic or the "blame game" response of equating failure with malfeasance.) But much of it surely resides within the institution of science itself and the culture surrounding it. As it has become more powerful and more entrenched (especially since World War II), it has gradually become more conservative, the fate of many successful institutions.

Breakthroughs often violate some accepted result. Which peer reviewer or grant program director or panel member is inclined to fund such a proposal when the violated item is entrenched as "true"? Which junior investigator will risk her career to submit such a proposal? Ironically, in many fields the only researchers who can do so fairly safely are established people beyond their prime of originality.

Scientific realism is the "textbook" model of science. It has been an indispensable model in many respects and a convenient way of presenting science to a relatively uninformed public. Unfortunately, however, it also presents a distorted conception of how science works, one that affects the education of future scientists, science education more generally, the resulting popular and political conceptions of how science works, and how many scientists regard their craft. It thereby contributes to the current conservatism of journal referees, grants panels, and administrators. Attention to what we might term "now-dead science" is certainly needed to gain problem-solving skills and to provide still-useful models, but why is so little attention given to what we don't know, to areas of disagreement or confusion as well (Firestein 2012; 2015)? Max Planck's professor, Philipp von Jolly, notoriously advised him not to go into physics on the ground that there was nothing interesting left to do. Fortunately, young Planck did not take the advice.

······

Notes to Chapter Eight

1. I have questioned some points in the Price-Rescher argument in Nickles 2013.
2. For a fuller account, see Nickles 2017 and forthcoming.
3. Weinberg 1992; Hawking 1998, 175. I agree with Philip Kitcher (this volume): "The idea of a 'theory of everything' is an absurd fantasy." Here and below, the reader will find resonances to Stanford 2006 as well as to Laudan 1981 and Fine 1986.

Literature Cited

Fine, Arthur. 1986. *The Shaky Game: Einstein, Realism, and the Quantum Theory*. Chicago: University of Chicago Press.

Firestein, Stuart. 2012. *Ignorance: How It Drives Science*. New York: Oxford University Press.

———. 2015. *Failure: Why Science Is So Successful*. New York: Oxford University Press.

Giere, Ronald. 2006. *Scientific Perspectivism*. Chicago: University of Chicago Press.

Hawking, Steven. 1988. *A Brief History of Time*. New York: Bantam Books.

Kahneman, Daniel. 2011. *Thinking Fast and Slow*. New York: Farrar, Straus and Giroux.

Kuhn, Thomas. (1962) 1970. *The Structure of Scientific Revolutions*. 2nd ed., containing "Postscript—1969." Chicago: University of Chicago Press.

Laudan, Larry. 1981. "A Confutation of Convergent Realism." *Philosophy of Science* 48: 19–49.

———. 1984. *Science and Values: The Aims of Science and Their Role in Scientific Debate*. Berkeley: University of California Press.

McIntyre, Lee. 2015. *Respecting Truth: Willful Ignorance in the Internet Age*. London: Routledge.

McMullin, Ernan. 1985. "Galilean Idealization." *Studies in History and Philosophy of Science Part A* 16 (3): 247–73.

Nickles, Thomas. 2006. "Heuristic Appraisal: Context of Discovery or Justification?" In *Revisiting Discovery and Justification: Historical and Philosophical Perspectives on the Context Distinction*, edited by Jutta Schickore and Friedrich Steinle, 159–82. Dordrecht, Netherlands: Springer.

———. 2013. "Creativity, Nonlinearity, and the Sustainability of Scientific Progress."

In *Creativity, Innovation, and Complexity in Science*, edited by Wenceslao González, 143–72. A Coruña, Spain: Netbiblio.

———. 2017. "Nonrealism: Objections and Replies." In *Varieties of Scientific Realism*, edited by Evandro Agazzid. Cham, Switzerland: Springer.

———. Forthcoming. "Do Cognitive Illusions Make Scientific Realism Deceptively Attractive?" In *New Approaches to Scientific Realism*, edited by Wenceslao González. Articles on scientific realism in volumes edited by Evandro Agazzi and Wenceslao González.

Oreskes, Naomi, and Erik Conway. 2010. *Merchants of Doubt*. New York: Bloomsbury Press.

Poincaré, Henri. 1952. *Science and Hypothesis*. New York: Dover. Originally published in French in 1905.

Popper, Karl. 1963. *Conjectures and Refutations: The Growth of Scientific Knowledge*. New York: Basic Books.

Price, Derek. (1963) 1968. *Little Science, Big Science*. 2nd ed. New York: Columbia University Press.

Quoidbach, Jordi, Daniel Gilbert, and Timothy Wilson. 2013. "The End of History Illusion." *Science* 339 (January 4): 96–98.

Rescher, Nicholas. 1978. *Scientific Progress*. Oxford: Blackwell.

———. 2006. *Epistemetrics*. Cambridge: Cambridge University Press.

Shapere, Dudley. 1984. *Reason and the Search for Knowledge*. Dordrecht, Netherlands: Reidel.

Smolin, Lee. 2006. *The Trouble with Physics: The Rise of String Theory, the Fall of a Science, and What Comes Next*. New York: Mariner Books.

Stanford, P. Kyle. 2006. *Exceeding Our Grasp: Science, History, and the Problem of Unconceived Alternatives*. Oxford: Oxford University Press.

Tierney, John. 2013. "Why You Won't Be the Person You Expect to Be." *New York Times*, January 4.

Turner, Frederick Jackson. 1920. *The Frontier in American History*. New York: Holt.

van Fraassen, Bas. 1980. *The Scientific Image*. Oxford: Oxford University Press.

———. 2001. "Constructive Empiricism Now." *Philosophical Studies* 106: 151–70.

Weinberg, Steven. 1992. *Dreams of a Final Theory: The Scientist's Search for the Ultimate Laws of Nature*. New York: Pantheon.

Wimsatt, William C. 2007. *Re-engineering Philosophy for Limited Beings*. Cambridge, MA: Harvard University Press.

Winsberg, Eric. 2010. *Science in the Age of Computer Simulation*. Chicago: University of Chicago Press.

Worrall, John. 1989. "Structural Realism: The Best of Both Worlds?" *Dialectica* 43: 99–124.

9

The Fundamental Argument
against Scientism

RIK PEELS

Introduction

Scientism is a view that, it seems, has recently become increasingly popular in science, philosophy, and especially popular science writing. The term *scientism* is used in a wide variety of ways, though. Some use it to refer to the thesis that only those things exist that can in principle be investigated by natural science. Others take it to be the claim that the humanities should adopt the methods of natural science or that they should even be replaced with natural science. Still others use it to refer to the thesis that science has no principled limits (e.g. Boudry, this volume). Boudry and Pigliucci, in their introduction to this volume, also point out that the term has a wide variety of meanings. (For a conceptual map that gives an overview of the main varieties of scientism, see Peels 2018 and Stenmark 2001.)

Most often, though, the term denotes the thesis that only natural science can reliably deliver knowledge or even mere rational belief, either in a specific realm or in general (for a similar understanding, see Blackford, Kitcher, and Law, this volume; for a defense, see Rosenberg, Kalef, this volume). Thus, we should mistrust the nonscientific sources of belief that we use in our daily lives, such as introspection and memory. According to adherents of this kind of scientism, we have good empirical reasons to think that nonscientific sources of belief are unreliable. (Elsewhere, I have discussed some of these empirical arguments in detail; see Peels 2016.) Hence, if we want to know and understand the world and ourselves, we should trust only natural science. This thesis is often referred to as "epistemological scientism." Henceforth, I use the word *scientism* as shorthand for "epistemological scientism."

This chapter presents and discusses a major worry for scientism. For reasons that will become clear shortly, I call my line of reasoning the *Fundamental Argument* against scientism. The argument is that the fundament of natural science itself consists of the deliverances of nonscientific sources of belief, such as auditory perception, memory, and logical intuition, so that if we claim that we should discard any nonscientific sources of belief, we make an untenable claim.[1] For example, astronomers cannot investigate supernova S Andromedae if they do not rely on their visual perceptions when they make observations through a telescope. A biologist investigating the rhinoceros hornbill, one of Borneo's many birds, in the rain forest will inevitably use her auditory perception. A chemist cannot continue her research on the molecular structure of a newfound substance if she does not rely on her memory of what she did yesterday or even a few seconds ago. And a mathematician working on a proof for the Riemann hypothesis can do her work only if she trusts her basic mathematical intuitions about calculation.

The aim of this chapter is to provide an answer to the question whether the Fundamental Argument holds water. In order to do so, I first spell out further what I mean by *scientism* (§2). Then, I lay out the Fundamental Argument in detail (§3). Subsequently, I discuss three objections that might be leveled against the Fundamental Argument. Since, as far as I know, this argument against scientism has not yet been discussed in the literature, I have selected these objections on the basis of conversations I had with people on this topic, and have tried to build the strongest case for them. The objections are, respectively, that scientism suffers merely from a local rather than a global fundamental problem (§4); that the results of natural science are verified, confirmed, or repeatable, whereas the deliverances of nonscientific belief sources are not (§5); and that the results of natural science are intersubjective, whereas the deliverances of nonscientific belief sources are not (§6). I argue that each of these objections is unconvincing. I then consider two more sophisticated versions of scientism that are meant to sidestep the Fundamental Argument. On Scientism 2.0, only natural science *and* those nonscientific sources of belief that are presupposed by natural science deliver knowledge

(§7). On Scientism 3.0, only those propositions can be known that can *in principle* be known on the basis of natural science. I argue that both of these more sophisticated versions of scientism either fail to escape the Fundamental Argument or become trivial and uninteresting claims (§8).

Scientism

I take scientism to be a thesis about natural science, such as biology, chemistry, and physics, since paradigm cases of scientism are theses that put the natural sciences center stage rather than, say, history or psychology.

As I indicated, I will focus on scientism as an *epistemological* rather than an ontological claim. This is how the term *scientism* is usually understood. According to Ian Barbour, for instance, scientism is the claim that "the scientific method is the only reliable path to knowledge" (1990, 4); according to Roger Trigg, it is the view that "science is our only means of access to reality" (1993, 90). (See also De Ridder 2014; Pigliucci 2013, 144.)

The epistemological focus, though, still leaves room for a wide variety of theses. Let me point out two distinctions that can be used to further specify which variety of scientism we aim to defend or criticize. First, scientism can be understood as the claim that only natural science (a) delivers rational or reasonable belief; (b) produces knowledge; or (c) reliably leads to rational belief or knowledge. These theses are conceptually distinct. We might think, for instance, that nonscientific beliefs can still be rational or reasonable, but that they cannot constitute knowledge. Or we might think that nonscientific sources of belief *incidentally* rather than *reliably* produce knowledge. It seems that (a) is the strongest variety, whereas (c) is the weakest. For, if nonscientific sources cannot even produce rational beliefs, then they surely cannot lead to knowledge or reliably lead to rational belief; and if nonscientific belief sources cannot produce knowledge, they cannot reliably lead to knowledge. I will mostly focus on those versions that say that only natural science delivers knowledge, for if nonscientific sources can also deliver knowledge, then they can deliver rational belief; and the examples I provide will give us good reason to think that they can do so reliably.

The second dimension along which varieties of scientism could be distinguished concerns the nonscientific sources of belief that are discarded. There is of course a wide variety of such sources: vision, taste, smell, hearing, and touch (various senses), memory, introspection, metaphysical intuition, logical intuition, mathematical intuition, linguistic intuition, and so forth. Stronger versions of scientism will discard all these nonscientific sources of belief, whereas weaker versions will discard only some of them. Thus, Otto Neurath (1987, 7–11) and James Ladyman, Don Ross and David Spurrett (2007, 1–65) discard metaphysical intuition as unreliable, and Daniel Dennett (1991; 2003) and Eric Schwitzgebel (2011) reject introspection as being untrustworthy. Others make a much more general claim. According to Alex Rosenberg, for instance, scientism

> is the conviction that the methods of science are the only reliable ways to secure knowledge of anything; that science's description of the world is correct in its fundamentals. . . . Science provides all the significant truths about reality, and knowing such truths is what real understanding is all about. . . . Being scientistic just means treating science as our exclusive guide to reality, to nature—both our own nature and everything else's. (Rosenberg 2011, 6–8)

(For a similar claim, see Atkins 1995.) In what follows, I focus on the stronger version of scientism that says that *in any domain* of reality, only natural science delivers knowledge. Below, we will see to which weaker thesis the adherent of scientism will have to retreat in order to defend a plausible position. We can thus define *scientism* as follows:

Scientism
Only natural science delivers knowledge.

This is a rather unsophisticated form of scientism that we often find in popular science writing. Below, I will discuss two more sophisticated versions of scientism. I nevertheless discuss *Scientism* in detail, because, as we shall see, the main problem with *Scientism* (laid out in the Fundamental Argument) returns in one way or another for the more developed versions of scientism.

The Fundamental Argument against Scientism

We find a rather short formulation of the Fundamental Argument in the writings of Mary Midgley:

> Science cannot stand alone. We cannot believe its propositions without first believing in a great many other . . . things, such as the existence of the

external world, the reliability of our senses, memory and informants, and the validity of logic. If we do believe in these things, we already have a world far wider than that of science. (Midgley 1992, 108)

Similar thoughts can be found in Stenmark 2001, 26–28. Even though the word *knowledge* is not actually used in this passage, the idea is clearly that if in doing science we do not assume the existence of many things that can be known even though they are in no way based on science, then science cannot deliver knowledge either. We find another brief characterization of the Fundamental Argument in an article by René van Woudenberg:

> Another response . . . might be to bite the bullet and deny that extra-scientific beliefs ever amount to knowledge. This, however, would be deeply problematic. For scientific knowledge depends in many ways on extra-scientific knowledge, for instance, on what we know through perception, such as that the thermometer now reads 118 degrees Fahrenheit. Without such extra-scientific knowledge it is hard to see how science could even get started. (van Woudenberg 2013, 26)

These quotations raise all sorts of questions, though. Exactly which sources of nonscientific belief are indispensable for doing science? Why should we think that this is the case? And precisely how is the argument supposed to run?

Let us first consider the structure of the argument and then return to the first and second questions. The argument can of course be structured in different ways, but one plausible way is to phrase is as a *reductio*:

(1) Only natural science delivers knowledge. [Scientism; Ass. for *reductio*]
(2) If only natural science delivers knowledge, then nonscientific sources of belief do not. [from (1)]
(3) Nonscientific sources of belief do not deliver knowledge. [from (1) & (2)]
(4) Natural science delivers knowledge. [Prem.]
(5) In doing natural science, scientists inevitably rely on nonscientific sources of belief. [Prem.]
(6) If, in doing science, natural scientists inevitably rely on nonscientific sources of belief, then the results of science are instances of knowledge only if those nonscientific sources of belief produce knowledge. [Prem.]
(7) Either natural science does *not* deliver knowledge, or nonscientific sources of belief *do* deliver knowledge. [from (5) & (6)]

(8) Thus, *either* natural science (reliably) delivers knowledge *and* it does
not, or nonscientific sources of belief deliver knowledge *and* they do
not. [contradiction; from (3), (4), (7)]

(9) Therefore, ~(1). [RAA][2]

Both (1) and (2) of this formalization of the Fundamental Argument are true
by definition, and (3) follows from (1) and (2) by *modus ponens*. Proposition (4) is a platitude. Premise (7) follows from (5) and (6) by disjunction-
conditional equivalence, and (8) follows from the conjunction of (3), (4), and
(7). Thus, the evidential weight is on the shoulders of (5) and (6).

We can see why (5) is plausible by imagining a situation in which we are
allowed to do science, but *not* to use any nonscientific sources of belief. It
should be obvious that in such a case we could *not* do science. We could not
use our memory and thus could not be sure whether it was *us* who gathered
the data on, say, the genetic profile of a sample of twins, and thus we would
not know whether those data have been gathered in a reliable way. We would
not be allowed to draw any inferences from the data, because in doing so
we would have to rely on basic logical intuitions, such as the intuition that
modus tollens is valid and an *ex consequentia* is not. We would not be allowed
to count the data, for in doing so we would have to rely on basic mathematical
intuitions, such as that $2 + 2 = 4$ and that $1 \times 3 = 3$. In fact, we would not be
allowed to *read* the data, for we would then use visual perception. We would
not even be allowed to believe that there *are* data, for that assumes that there
is an external world, and that intuition, surely, is a metaphysical nonscientific
intuition rather than a product of scientific inquiry.

What about introspection? Do scientists have to rely on introspection in
doing scientific inquiry? According to influential approaches in contemporary psychology and neuroscience, such as so-called neurophenomenology,
front-loaded phenomenology, and descriptive experience sampling, these
fields are at least severely impoverished if not allowed to base their research
at least partly on introspection—see, respectively, Jack and Shallice 2001; Jack
and Roepstorff 2003; Lutz and Thompson 2003; Gallagher 2003; Hurlburt
and Heavey 2001. Introspection might even be necessary for other natural
sciences. According to many philosophers, we know by way of introspection
what we believe and what we intend. But knowing what we believe about,
say, some experiment or knowing what we intend to do next is of course indispensable for natural scientific research. Thus, we have good reason to think
that even introspection, as a source of belief, is necessary for natural science.

That leaves us with premise (6). On this premise, if, in doing science, scientists inevitably rely on nonscientific belief sources, then the results of science count as knowledge only if those nonscientific sources produce knowledge. This premise is supported by, for instance, the contemporary literature on Gettier cases.[3] Imagine, for example, that there is a dog in the field in front of me, but that I take it to be a sheep and, therefore, believe that there is a sheep in the field. My belief that there is a sheep in the field, though, happens to be true, because, unbeknownst to me, there is a sheep behind a large rock in the field. In that case, I do not *know* that there is a sheep in the field, even though my belief is true (for this example, see Chisholm 1966, 105). Thus, it is hard to see how a belief can constitute knowledge if it is based on a belief that fails to be an instance of knowledge.[4] This is true not only for knowledge but also for rationality. If I irrationally believe that Alpha-Centaurs have invaded our planet and, knowing that if they exist, Alpha-Centaurs are aliens, infer that aliens have invaded our planet, then that does not make my belief that aliens have invaded our planet rational, even though the deduction is perfectly valid. Beliefs cannot be rational if they are based on beliefs that are irrational. There is good reason, then, to think that (6) is true.

Since it seems hard to deny that scientists in doing scientific research inevitably use, say, their senses, all three objections to the Fundamental Argument that I discuss are directed against premise (6), which says that if, in doing science, natural scientists inevitably rely on nonscientific sources of belief, then the results of science are known only if those nonscientific sources produce knowledge.

I would like to stress that the Fundamental Argument does *not* presuppose *foundationalism*, which says that all knowledge is, in the end, based on beliefs that are not themselves based on other beliefs but that do constitute knowledge (see, for instance, Plantinga 1993). It is equally compatible with *coherentism*, which says that a true belief constitutes knowledge when it coheres in the right way with the other beliefs the cognitive subject in question has (see for example BonJour 1985; Lehrer 1990). Of course, the Fundamental Argument would look slightly different on coherentism. Most important, instead of (6) it would say something like this:

(6')
If, in doing science, natural scientists inevitably rely on nonscientific sources of belief, then: if beliefs from *scientific* sources A and B cohere *insufficiently* with beliefs from *nonscientific* sources C and D, then beliefs from both A and

B and C and D do *not* provide knowledge, and if they *do* cohere sufficiently, they *do* provide knowledge.

Basically, the idea here is that if rationality or knowledge is to be cashed out in terms of coherence and if scientists in doing science inevitably rely on nonscientific beliefs, then there are two options. On the one hand, if the relevant scientific beliefs cohere sufficiently with the relevant nonscientific beliefs, then, on coherentism, they *both* count as rational or as instances of knowledge, which would undermine scientism as I defined it, for then all sorts of nonscientific beliefs would also be rational and count as knowledge. On the other hand, if the relevant scientific beliefs do *not* cohere sufficiently with the relevant nonscientific beliefs, then, on coherentism, *both* do *not* count as rational or as instances of knowledge, and that would clearly contradict the spirit of scientism—science, after all, is supposed to deliver rational beliefs and knowledge. Thus, either way, scientism, on coherentism as well, faces the Fundamental Argument.

For the sake of clarity and simplicity, in this chapter I confine myself to that version of the Fundamental Argument that is cashed out in terms that are often used by foundationalists, even though it could equally well be spelled out in coherentist terms.

First Objection: The Fundamental Problem Is Only Local

The first objection is that the problem from which the Fundamental Argument is derived is only *local*, that is, confined to a particular realm in science, since we can correct certain beliefs in a particular scientific realm by using scientific research from another scientific realm. In scientific research, we could then turn from one scientific source of belief or scientific method—that might be individuated on different levels—to another, each time using the beliefs from other scientific sources. In that way, we might acquire knowledge that each of those scientific sources all by themselves would not have provided.

A guiding image here could be one that is well known from metaphysics, namely Theseus's ship. One sets sail across the ocean on a ship that has on board exactly the kind and number of wooden boards the ship consists of. Every day, one replaces one board of the ship, until most of the ship or even the entire ship has been replaced with new material. We can even assume that all the old boards have been thrown overboard. We would then, according to many, end up with an entirely new ship. Similarly, we might think, we can replace our nonscientific beliefs on the basis of scientific research while using other nonscientific beliefs, until we end up with knowledge, discarding all beliefs that are based on nonscientific sources of belief. Otto Neurath presents an idea along these lines:

> We are like sailors who on the open sea must reconstruct their ship but are never able to start afresh from the bottom. Where a beam is taken away a new one must at once be put there, and for this the rest of the ship is used as support. In this way, by using the old beams and driftwood the ship can be shaped entirely anew, but only by gradual reconstruction. (Neurath 1973, 199)

Well-known examples of such belief replacement can be found in research on so-called cognitive biases. Baumeister and Bushman have shown that people use a so-called representativeness heuristic. This means that they have a tendency to judge the likelihood or frequency of an occurrence by the extent to which the event resembles the typical case (for more on this bias, see Baumeister and Bushman 2010, 141). Well-known mistakes in statistical reasoning, such as the Prosecutor's Fallacy, which is committed, roughly, when one assumes that the probability of a random match is identical to the probability that the defendant is guilty, would be other examples (for a detailed discussion of the Prosecutor's Fallacy, see Thompson and Schumann 1987).

This argument points to an important phenomenon, to wit, correction of beliefs from one source by beliefs from other sources. Yet for three reasons, I do *not* think that it provides us with good reason to reject premise (6) of the Fundamental Argument. First, there has indeed been belief correction on the basis of scientific inquiry, but the percentage of those beliefs compared to all nonscientific beliefs that scientific inquiry is based on is negligible. Virtually all the scientists' beliefs based on visual perception, memory, introspection, basic logical intuitions, and so forth have never been subjected to revision on the basis of scientific inquiry. Thus, the fact that there has been belief correction on the basis of scientific research does not count against the thesis that if nonscientific sources of belief generally do not produce knowledge, then natural science does not do so either.

Second, it seems that we find a similar kind of belief correction from other belief sources in nonscientific contexts, even in everyday life. We think we poured ourselves some coffee, but as soon as we taste it, we realize that it is tea. We thought we put the book on our desk, but when we search for it, it turns out to be on one of the shelves. We even make corrections from the same belief source, sometimes from a more fine-grained faculty, sometimes from the same faculty. Thus, we think that we see a snake, but when we look again, we see it is merely a tree branch. The belief correction process, then, is not unique to the scientific context. On the contrary, this phenomenon often takes place with beliefs from nonscientific sources.

Third and most important, what the image of Theseus's ship fails to capture is that rationality and warrant—that which closes the gap between true belief and knowledge—or the lack thereof is transferred from one belief to another if we base one belief on another. In the same way as we cannot set sail without a boat, we cannot revise certain nonscientific beliefs without starting from certain nonscientific beliefs. Of course, we might later revise *those* nonscientific beliefs, but even if we do so, they will have influenced our other beliefs at the earlier occasion, given the fact that we cannot start from scratch, that is, without beliefs. If a board of a ship is replaced with a good board and is surrounded by other good boards, everything is fine. But if a belief B is based on belief B*, and if B* has a negative epistemic status in that it is irrational or lacks warrant, and if belief B* is later removed, then, if B is not based on other beliefs, it inherits the irrationality and lack of warrant of B*. Hence, if nonscientific sources of belief fail to deliver knowledge, we cannot, through a process of scientific inquiry, (reliably) arrive at knowledge; for, if nonscientific is indeed untrustworthy, we will have to work with beliefs from sources that are untrustworthy.

Second Objection: Scientific Results Are Confirmed or Repeatable

Another objection to the Fundamental Argument is that scientific results have, time and again, been confirmed, while our ordinary nonscientific beliefs have not. Now, it is of course highly controversial that science aims at or should aim at confirmation. Karl Popper (1968) famously argued that science should aim at the falsification of its hypotheses—or, at least, that it should formulate falsifiable hypotheses and set up experiments that could falsify those hypotheses. After all, pseudoscience, such as Alfred Adler's inferiority complex theory, has also been confirmed time and again. Popper's view that science should aim at falsification has itself also been seriously criticized, though; see Haack 2007, 34–37. Let us therefore assume that science aims at confirmation. Even if this is correct, this second objection to the Fundamental Argument fails for at least two reasons.

First, many of our ordinary nonscientific beliefs *have* been confirmed rather frequently. Every time I get home, I see that my couch is in my living room, that it is red, that it feels soft, and so forth. Friends and family that have visited my place have had the same sense experience. It is not clear that we can speak of "verification" in the case of logic, but I have considered and used *modus ponens* several times in my life, and unsurprisingly, it has never turned out to be invalid (it has never been falsified).

Second, it seems that if the nonscientific beliefs on which confirmed scientific hypotheses are based were irrational or lacked warrant, the confirmation of those scientific hypotheses would be of little epistemic value. Thus, the objection still assumes that at least in most cases (for most people), nonscientific beliefs constitute knowledge or, at least, are rational. In response to this, we might suggest that the very fact that, in the case of scientifically confirmed hypotheses, the same results were found among different individuals at different times makes those hypotheses somehow more reliable, and that scientific beliefs are therefore rational and count as knowledge, whereas (many) nonscientific beliefs do not. Since the objection has now switched from verification to intersubjectivity, I address it as a separate objection in the next section.

Alternatively, we could claim that the experiments on which our scientific theories are based can be repeated, whereas nonscientific beliefs provide us with no such opportunity. So here the idea is not so much that scientific theories have been confirmed or verified, but that *we can actually check* the results of science by simply *repeating* the experiment in question. For instance, if you question the law of Boyle, you can set up an experiment and thereby check for yourself whether $PV = k$, and whether $P_1V_1 = P_2V_2$. And if you want to check whether the first fundamental theorem of calculus, which says that an indefinite integral of a function can be reversed by differentiation, is correct, you can simply go through the proofs or, if you are good at math, come up with a proof yourself. If, however, someone remembers that she had a grapefruit for breakfast a week ago or introspects that she believes *right now, at time t* that Jerusalem was conquered by the Romans in 70 AD, there is no experiment, experience, or circumstance she can repeat or re-create to check whether those beliefs are indeed true. This is of course not to deny that she might again eat a grapefruit. However, the belief formed on the basis of that experience would differ from her original belief. And it is also not to deny that she might check whether the Romans did indeed conquer Jerusalem in 70 AD. That, however, will not tell her anything about her beliefs at *t*.

This objection suffers from at least two serious problems. First, to the extent that nonscientific cognition is concerned with general truths—in the same way that empirical science is concerned with general truths, such as monogenetic evolution or the law of gravity—there is good reason to think that the experiences and circumstances on which general nonscientific beliefs are based can be repeated. I believe that human beings are normally gifted with the ability to walk. That belief is not based on scientific research. But I can easily repeat the kind of experiences or circumstances on which that general belief is based by taking a walk to the city center. I hold the general belief that breaking an arm hurts. I have seen several cases in which people broke an arm, and it seemed to hurt really badly. I could repeat those circumstances by visiting a hospital or going to such extremes as breaking my own arm (true, I will not do that, but I think I *could* do it).

Second, to the extent that scientific experiments are repeatable and to the extent that that is a strong indication of their truth, that quality actually *entails* that nonscientific beliefs are rational and normally, if true, count as knowledge. After all, in repeating the experiments, we *again* use visual perception, auditory perception, memory, logical reasoning, and so forth to establish the results in question. If these nonscientific sources of belief did not produce knowledge, we would not get the same results time and again, or if we did, repeatability of the experiment or experience on which the belief is based is not a good criterion for that belief's constituting knowledge. Hence, to the extent that scientific results are repeatable and this tells us something about their truth, it actually *confirms* rather than *disproves* the rationality of nonscientific beliefs and thus counts against scientism.

Third Objection: Scientific Results Are Intersubjective

A third objection leveled against the Fundamental Argument says that in opposition to nonscientific beliefs, scientific results are *intersubjective*: several, often even many, cognitive subjects share scientific beliefs. When I see a squirrel in the forest or have a memory of a particular event at my primary school, that is *my* perception and *my* memory and, in many cases, *only* my perception and my memory. But when a scientist believes the second law of thermodynamics, that belief is shared by thousands of physicists around the world. Thus, even though natural science is *based on* nonscientific beliefs, such as individual scientists' observational beliefs, the results of science, such as scientific theories that are backed up by strong evidence, are widely shared, and therefore can count as knowledge and can be rationally embraced.

This argument is problematic for several reasons. Let me mention two of them. First, there are many things that we know intersubjectively, but that are in no way based on scientific research. I believe that soccer is played with a ball, that the Sun comes up in the east, and that it is generally warmer in Italy than it is in Norway, and so do millions of other people, but these beliefs are not based on scientific research. The argument, then, falsely suggests that scientific beliefs are intersubjective, whereas nonscientific beliefs are not. Of course, some nonscientific sources are—sometimes maybe even necessarily, as in the case of introspection—available only to the individual, but many nonscientific beliefs are intersubjective.

Second, it is not at all clear why we should think that intersubjective beliefs constitute knowledge, whereas beliefs from a source that is restricted to the individual do not. Consider beliefs based on introspection. Of course, we can be mistaken about things we take ourselves to know. As empirical research shows, especially in the area of knowing the reasons for our actions, there is room for error: we sometimes believe that we performed a particular action for reason X while what really motivated us is something different, Y. A landmark article on this issue is Nisbett and Wilson 1977. However, it is obvious that there are many things we know on the basis of introspection. To take a few examples from the beliefs I myself hold on the basis of introspection: I believe that it seems to me that there is a computer in front of me, I believe that I am now thinking about scientism, and I believe that I intend to finish a draft of this chapter today. It seems hard to deny that I *know* these things by way of introspection. But then the principle that we can know something only if it is believed intersubjectively is simply false.

Scientism 2.0: Natural Science and What It Presupposes

I have argued that the main objections to the Fundamental Argument against scientism are unconvincing. One might suggest, though, that *Scientism* can be revised in such a way that it meets the Fundamental Argument. First, one could suggest that we accept the deliverances of natural science *and* those nonscientific cognitive faculties the use of which are necessary for doing natural science. Thus, we can in general accept the deliverances of visual perception, memory, smell, and logical reasoning, because natural science is based on these beliefs; and we must thus presuppose that they constitute knowledge or, at least, can be held rationally:

Scientism 2.0
Only natural science and those nonscientific sources of belief that are presupposed by natural science deliver knowledge.

It is clear that the Fundamental Argument does not spell trouble for *Scientism 2.0*, since the latter wholeheartedly embraces most nonscientific sources of belief. Yet, *Scientism 2.0* is problematic for at least three other reasons.

First, scientism has now become so weak that we might wonder whether it still counts as a genuine variety of scientism. After all, on *Scientism 2.0*, it is perfectly legitimate to hold all sorts of beliefs that are not in any way based on science. If certain contemporary scientists are right, even introspection will have to be used in order to do good, say, neurophysiological research—as I pointed out above, approaches like neurophenomenology, front-loaded phenomenology, and descriptive experience sampling have gained popularity recently, and each of these relies on introspection.

Second, *Scientism 2.0* fails to ascribe a special status to beliefs based on scientific research, and many would therefore reject it as a variety of scientism. Of course, *Scientism 2.0* is still phrased in terms of science, but it actually amounts to a view on which countless beliefs—maybe even the vast majority of our beliefs—are rational and count as knowledge, even though they are not in any way the product of science or part of scientific research.

Third, if it is perfectly legitimate from an epistemic point of view to use memory, logical intuition, visual perception, and other nonscientific sources of belief because those beliefs constitute knowledge, even though these sources are far from infallible, then why should we exclude, say, many of our metaphysical intuitions? Given the problems that plague the three objections that I discussed in the three previous sections, the reply that these sources are not presupposed by science does, as such, not seem to count against them—we saw that objections to the Fundamental Argument from (the lack of) confirmation or (the lack of) intersubjectivity are unconvincing. This is not to deny that there may be a good argument to accept *most* nonscientific sources of belief and exclude *some*. We might argue, for instance, that there is more of a consensus on the deliverances of visual perception and memory than on morality. This is *not* obviously the case, though: scientism will need to be backed up by substantial argumentation if this is the position advocated.

Let me give two examples to illustrate this point. First, people sometimes greatly vary on how they remember certain events. Memories of exactly what happened in a traffic accident, even from those not directly involved, are well known to differ greatly among witnesses. Second, sociological research has shown that a wide variety of basic moral convictions, such as that human life has value and that raping one's children is morally wrong, are widely shared among different cultures. Maybe there is an argument to be made here, but as these examples show, it will require substantial reasoning to provide a plausible demarcation criterion to distinguish which nonscientific sources are to be allowed in (as sources of rational belief) and which are not.

This is not to say that there cannot be good arguments to ascribe a low epistemic status to beliefs from a specific nonscientific doxastic source, such as moral intuition or religious experience. However, arguments along these lines, such as *debunking explanations* of moral realist beliefs and religious beliefs, are highly controversial as things stand (see, for instance, Bering 2011; Clark and Barrett 2010). Future empirical research and philosophical evaluation will have to show whether these arguments hold water.

Scientism 3.0: What Can in Principle Be Known by Natural Science

Another route we could take in response to the Fundamental Argument is to maintain the priority of natural science, while leaving room for the thought that nonscientific sources of belief often produce rational belief or knowledge, by claiming that if something can be known, then it is either the result of natural science or something that could *in principle* be known by applying the methods of natural science to it:

Scientism 3.0
Only those propositions can be rationally believed or known that can *in principle* be rationally believed or known on the basis of natural science.

It is not unusual to understand scientism along these lines. This is also Michael Peterson's understanding of the word *scientism* when he defines it as the idea that "science tells us everything there is to know about what reality consists of" (Peterson et al. 1991, 36). And René van Woudenberg defines *scientism* as "the view that all truths can be known via scientific inquiry" (2011, 185). *Scientism 3.0* is not a straw man. Bertrand Russell seems to advocate a position like *Scientism 3.0* when he says, "Whatever can be known, can be known by means of science" (1946, 863). And Rudolph Carnap is even more explicit when he says, "The total range of life still has many other dimensions outside of science . . . within its dimension, science meets no barrier. . . . When we say that scientific knowledge is unlimited, we mean: *there is no question whose answer is in principle unattainable by science*" (1967, 290; italics are from Carnap).

However, *Scientism 3.0* is defective in at least three regards. First, it is not at all clear what we mean when we say that something can *in principle* be known by way of natural science. It cannot mean that it can be known by science as it is currently done, because science is frequently defective, faulty, and sloppy, and it is likely that we are currently working with a significant number of false hypotheses. Also, given the track record of science, it seems highly plausible that there are all sorts of things that can be known, but that we cannot yet know by science, because science has not yet developed enough. So maybe "in principle" means something like: "perfectly performed and in its best future state." But who knows what can be known by future science, perfectly performed? Maybe moral and religious truths are among them, if there are indeed such truths. However, if that is the case, then it cannot be ruled out that we can know and rationally believe such religious and moral propositions now already, for that would be compatible with *Scientism 3.0*.[5] But it surely goes against the spirit of scientism to say that for all we know, we might now rationally believe and know all sorts of moral and religious propositions that are currently without any support from natural science.

Second, there are all sorts of things we *know*, even though we *cannot* know them on the basis of scientific research, not even in principle. I know that I feel dizzy, I know that I have a headache, I know that I put fire to a tree in the garden of our neighbor when I was eight years old. This is not to deny that science at some point may be able to read people's states of mind. It seems, however, that *if* science is ever able to do so, it will be able to do so because people *report* feeling dizzy, having a headache, or having that particular memory, and researchers can consequently relate people's reports to certain brain patterns. If we will ever be able to know these things on the basis of scientific research, it will therefore be because people *know* such things as whether they feel dizzy and *report* to scientists or others involved in the experiment that they feel dizzy. Only once people have reported that they feel dizzy, based on their knowledge of the fact that they feel dizzy, can scientists relate feeling dizzy to certain brain states and consequently reliably predict on the basis of the results from brain scanners whether a particular person feels dizzy.

Third, *Scientism 3.0* does not escape the Fundamental Argument. Take *modus ponens*, elementary mathematical truths, or statements about what criteria a theory should meet in order to count as a good scientific theory, such as simplicity and explanatory power. These are *not* propositions for which we have good scientific evidence. Rather, they are convictions on which science is based. (For more examples along these lines, see van Woudenberg 2011, 182–85; 2013, 26.) Thus, even though a significant number of propositions that most people take themselves to know from nonscientific sources can be known by scientific research, it seems that there is still a wide variety of propositions that are known and rationally believed by way of nonscientific cognition, but that cannot be known or rationally believed by way of natural scientific research. Hence, we should reject not only *Scientism 2.0* but *Scientism 3.0* as well.

Conclusion

I conclude that every claim to the effect that only natural science can (reliably) deliver rational belief or knowledge is untenable, because, as the Fundamental Argument shows, it is self-defeating. As soon as scientism discards nonscientific sources of belief, it removes its own foundations and thus collapses. If scientism is to be plausible, it should make a significantly more modest claim, such as the claim that a *specific* nonscientific source of belief, such as belief formation about one's reasons for performing a past action, is insufficiently reliable to count as knowledge; or the claim that natural science *is more reliable* in leading us to knowledge than some of our nonscientific sources of belief; or that the deliverances of natural science are *more rational* to believe than the deliverances of some of our nonscientific sources of knowledge. Such a more modest claim may very well not suffer from the Fundamental Argument. It will call for a more careful scrutiny of other issues, though, such as the extent to which the track record of science warrants the claim that the deliverances of natural science are more rational than those of the nonscientific sources of belief. Thus, even if it is more modest, whether it is more plausible remains to be seen. I leave such an assessment for another occasion.

Acknowledgments

For their helpful comments on previous versions of this chapter, I thank the editors of this volume, Maarten Boudry and Massimo Pigliucci; for inspiring discussions on scientism, I thank the attendees of a workshop the editors organized at the City University of New York, including Carol Cleland, Taner Edis, Noretta Koertge, Thomas Nickles, Jesse Prinz, Don Ross, and Mariam Thalos. I would also like to thank Lieke Asma, Terence Cuneo, Leon de Bruin, Jeroen de Ridder, Naomi Kloosterboer, Kelvin McQueen, Scott Robins, Emanuel Rutten, Russ Shafer-Landau, Gijsbert van den Brink, Pieter van der Kolk, Hans van Eyghen, René van Woudenberg, Sander Verhaegh, and Albert Visser. Finally, I am grateful to the anonymous reviewer for University of Chicago Press for insightful comments on an earlier version of this chapter.

Publication was made possible through the support of a grant from the Templeton World Charity Foundation. The opinions expressed herein are those of the author and do not necessarily reflect the views of the Templeton World Charity Foundation.

••••••

Notes to Chapter Nine

1. The main thesis of this chapter should *not* be confused with the thesis that scientism is self-referentially incoherent, that is, that scientism is untenable because scientism itself is not supported by science. I have discussed that claim elsewhere. See Peels 2017. Here, my point is that if scientism is true, it will have to discard science as well, since science is based on nonscientific beliefs.

2. Just to be explicit: premise (6) should be read as (5) → ((4) → ~(3)), premise (7) as ~(4) v ~(3), and premise (8) as ((4) & ~(4)) v (~(3) & (3)).

3. This is even the case in the original Gettier examples. Smith falsely believes and hence does not know that Jones will get the job, and Smith falsely believes and hence does not know that Jones owns a Ford. See Gettier 1963, 121–23.

4. Note that I am *not* setting up myself for an infinite regress here; the claim is merely that *to the extent that a belief is based on other beliefs*, the former counts as knowledge only if the latter also count as knowledge. Many instances of knowledge will not be based on beliefs that count as knowledge because they are not based on any other beliefs at all.

5. Of course, what I say here strongly resembles Hempel's (1969) dilemma regarding what we should consider to be physical: that which is physical on our current scientific theories or that which ideal future science tells us is physical.

Literature Cited

Atkins, Peter W. 1995. "Science as Truth." *History of the Human Sciences* 8 (2): 97–102.

Barbour, Ian G. 1990. *Religion in an Age of Science: The Gifford Lectures 1989–1991*. Vol. 1. New York: SCM Press.

Baumeister, Roy F., and Brad J. Bushman. 2010. *Social Psychology and Human Nature*. Belmont, CA: Wadsworth.

Bering, Jesse. 2011. *The God Instinct: The Psychology of Souls, Destiny, and the Meaning of Life*. New York: W. W. Norton.

BonJour, Laurence. 1985. *The Structure of Empirical Knowledge*. Cambridge, MA: Harvard University Press.

Carnap, Rudolf. 1967. *The Logical Structure of the World: And Pseudoproblems in Philosophy*. Translated by Rolf A George. Chicago: Open Court.

Chisholm, Roderick M. 1966. *Theory of Knowledge*. Englewood Cliffs, NJ: Prentice Hall.

Clark, Kelly James, and Justin L. Barrett. 2010. "Reformed Epistemology and the Cognitive Science of Religion." *Faith and Philosophy* 27 (2): 174–89.

Dennett, Daniel C. 1991. *Consciousness Explained*. London: Penguin Press.

———. 2003. "Who's On First? Heterophenomenology Explained." *Journal of Consciousness Studies* 10 (9–10): 19–30.

De Ridder, Jeroen. 2014. "Science and Scientism in Popular Science Writing." *Social Epistemology Review and Reply Collective* 3 (12): 23–39.

Gallagher, Shaun. 2003. "Phenomenology and Experimental Design: Toward a Phenomenologically Enlightened Experimental Science." *Journal of Consciousness Studies* 10 (9–10): 85–99.

Gettier, Edmund L. 1963. "Is Justified True Belief Knowledge?" *Analysis* 23 (6): 121–23.

Haack, Susan. 2007. *Defending Science—within Reason: Between Scientism and Cynicism*. Amherst, NY: Prometheus Books.

Hempel, Carl Gustav. 1969. "Reduction: Ontological and Linguistic Facets." In *Philosophy, Science, and Method: Essays in Honor of Ernest Nagel*, edited by Sidney Morgenbesser, Patrick Suppes, and Morton White, 179–99. London: MacMillan.

Hurlburt, Russell T., and Christopher L. Heavey. 2001. "Telling What We Know: Describing Inner Experience." *Trends in Cognitive Sciences* 5 (9): 400–403.

Jack, Anthony I., and Andreas Roepstorff. 2003. "Editorial Introduction: Why Trust the Subject?" In *Trusting the Subject: The Use of Introspective Evidence in Cognitive Science*, vol. 1 of a special series. *Journal of Consciousness Studies* 10 (9–10): v–xx.

Jack, Anthony I., and Tim Shallice. 2001. "Introspective Physicalism as an Approach to the Science of Consciousness." *Cognition* 79 (1–2): 161–96.

Ross, Don, James Ladyman, and David Spurrett. "In Defence of Scientism." In *Every Thing Must Go: Metaphysics Naturalized*, by James Ladyman and Don Ross, 1–65. Oxford: Oxford University Press.

Lehrer, Keith. 1990. *Theory of Knowledge*. Boulder, CO: Westview Press.

Lutz, Antoine, and Evan Thompson. 2003. "Neurophenomenology: Integrating Subjective Experience and Brain Dynamics in the Neuroscience of Consciousness." *Journal of Consciousness Studies* 10 (9–10): 31–52.

Midgley, Mary. 1992. *Science as Salvation*. London: Routledge.

Neurath, Otto. 1973. "Anti-Spengler." In *Empiricism and Sociology*, edited by Marie Neurath and Robert S. Cohen, 158–213. Dordrecht, Netherlands: Reidel.

———. 1987. "Unified Science and Psychology." In *Unified Science*, edited by Brian McGuinness, 1–23. Dordrecht, Netherlands: Kluwer.

Nisbett, Richard E., and Timothy DeCamp Wilson. 1977. "Telling More than We Can Know: Verbal Reports on Mental Processes." *Psychological Review* 84 (3): 231–59.

Peels, Rik. 2016. "The Empirical Case against Introspection." *Philosophical Studies* 173 (9): 2461–85.

———. 2017. "Scientism and the Argument from Self-Referential Incoherence." Unpublished manuscript. PDF.

———. 2018. "A Conceptual Map of Scientism." In *Scientism: A Philosophical Exposition and Evaluation,* edited by Jeroen de Ridder, Rik Peels, and René van Woudenberg. New York: Oxford University Press.

Peterson, Michael, William Hasker, Bruce Reichenbach, and David Basinger. 1991. *Reason and Religious Belief.* Oxford: Oxford University Press.

Pigliucci, Massimo. 2013. "New Atheism and the Scientistic Turn in the Atheism Movement." *Midwest Studies in Philosophy* 37 (1): 142–53.

Plantinga, Alvin. 1993. *Warrant and Proper Function.* New York: Oxford University Press.

Popper, Karl R. 1968. *The Logic of Scientific Discovery.* 3rd ed. London: Hutchinson.

Rosenberg, Alex. 2011. *The Atheist's Guide to Reality: Enjoying Life without Illusions.* New York: W. W. Norton.

Russell, Bertrand. 1946. *History of Western Philosophy, and Its Connection with Political and Social Circumstances from the Earliest Times to the Present Day.* London: Allen and Unwin.

Schwitzgebel, Eric. 2011. *Perplexities of Consciousness.* Cambridge, MA: MIT Press.

Stenmark, Mikael. 2001. *Scientism: Science, Ethics and Religion.* Aldershot, UK: Ashgate.

Thompson, William C., and Edward L. Schumann. 1987. "Interpretation of Statistical Evidence in Criminal Trials: The Prosecutor's Fallacy and the Defense Attorney's Fallacy." *Law and Human Behavior* 11 (3): 167–87.

Trigg, Roger. 1993. *Rationality and Science.* Oxford: Blackwell.

Woudenberg, René van. 2011. "Truths that Science Cannot Touch." *Philosophia Reformata* 76 (2): 169–86.

———. 2013. "Limits of Science and the Christian Faith." *Perspectives on Science and Christian Faith* 65 (1): 24–36.

10

Scientism and Pseudoscience: In Defense of Demarcation Projects

MASSIMO PIGLIUCCI

Prologue: Popper and the Demarcation Problem

Ever since Popper (1963), "demarcation projects" in philosophy of science are concerned with understanding what, if anything, separates some general kinds of epistemic practices. The original demarcation problem, the one Popper himself was interested in, focused on the difference between science and pseudoscience. He famously wrote that Einstein's theory of general relativity—which had recently[1] been spectacularly confirmed during the 1919 total eclipse of the Sun—was a paradigmatic example of good science, while astrology, Marxist theories of history, or Freudian and Adlerian psycho-analysis, were equally paradigmatic examples of pseudoscience. Here is how he put it while explaining his famous criterion of falsifiability to distinguish between science and pseudoscience:

> Einstein's theory of gravitation clearly satisfied the criterion of falsifiability. Even if our measuring instruments at the time did not allow us to pronounce on the results of the tests with complete assurance, there was clearly a possibility of refuting the theory.

Astrology did not pass the test. Astrologers were greatly impressed, and misled, by what they believed to be confirming evidence—so much so that they were quite unimpressed by any unfavourable evidence. Moreover, by making their interpretations and prophecies sufficiently vague they were able to explain away anything that might have been a refutation of the theory had the theory and the prophecies been more precise. In order to escape falsification they destroyed the testability of their theory. It is a typical soothsayer's trick to predict things so vaguely that the predictions can hardly fail: that they become irrefutable.

The Marxist theory of history, in spite of the serious efforts of some of its founders and followers, ultimately adopted this soothsaying practice. In some of its earlier formulations . . . their predictions were testable, and in fact falsified. Yet instead of accepting the refutations the followers of Marx re-interpreted both the theory and the evidence in order to make them agree. In this way they rescued the theory from refutation. . . . They thus gave a "conventionalist twist" to the theory; and by this stratagem they destroyed its much advertised claim to scientific status.

The two psycho-analytic theories were in a different class. They were simply non-testable, irrefutable. There was no conceivable human behavior which could contradict them. . . . I personally do not doubt that much of what they say is of considerable importance, and may well play its part one day in a psychological science which is testable. But it does mean that those "clinical observations" which analysts naively believe confirm their theory cannot do this any more than the daily confirmations which astrologers find in their practice. (Popper 1963, 36–37)

Popper changed his mind about a number of issues related to falsification and demarcation. For instance, he initially resisted a strong role for verification in establishing scientific theories, but later on agreed that verification—especially of very daring and novel predictions—is part of a sound scientific approach. He also changed his mind about the status of Marxist theories of history (and about evolutionary theory as well: Popper 1978), conceding that even the best scientific theories are often somewhat shielded from falsification because of their connection to ancillary hypotheses and background assumptions (the so-called Duhem-Quine theses: Ariew 1984). However, he persisted in considering Marxist historicism a pseudoscience, because in his opinion Marxists engaged in ad hoc rescues, similar to those practiced by astrologers (Thornton 2013).

Nonetheless, the basic point is that Popper single-handedly put on the map a problem for philosophers of science that is still with us today: how to make sense of the differences between science and pseudoscience. This lasted until 1983, when Larry Laudan published a highly influential paper, "The Demise of the Demarcation Problem" (1983), in which he argued that demarcation projects are a waste of time for philosophers, since—among other reasons—it is unlikely to the highest degree that anyone will ever be able to come up with small sets of necessary and jointly sufficient conditions to define *science, pseudoscience,* and the like. And without such sets, Laudan argued, the quest for any principled distinction between those activities is hopelessly quixotic.

A number of authors have more recently maintained that Laudan was too quick to dismiss the demarcation problem (Pigliucci and Boudry 2013), and that pathways forward are available if we abandon the strict requirement for necessary and jointly sufficient conditions, treating science, pseudoscience, and so forth as Wittgensteinian "family resemblance" concepts instead (Pigliucci 2013). That is because surely one gives up too much, as a philosopher, if one rejects the notion that there is something fundamentally different between, say, astrology and astronomy. And if philosophers cannot make sense of *that* obvious distinction, then they might as well pack up and devote themselves to less intellectually demanding activities.

All the above are pertinent to discussions of scientism because, in a sense, they mirror discussions about pseudoscience: if pseudosciences are nonscientific activities that mask themselves as genuine sciences, scientism—at the least under some interpretations of the term—is a scientific activity that projects itself into domains or areas of inquiry where it does not (allegedly) properly belong. Just as an important aspect of Popper's demarcation problem required a better understanding of what counts as science and as pseudoscience, so the scientism problem requires a better understanding of what counts as science and as the much broader category of nonscience.

What Is Scientism?

I like to approach complex problems by starting simply, going back to the basics. In this case, let us begin with a standard dictionary definition of *scientism. Merriam-Webster's Collegiate Dictionary* characterizes it as "an exaggerated trust in the efficacy of the methods of natural science applied to all areas of investigation (as in philosophy, the social sciences, and the humanities)." To date, and despite reams of paper and much electronic ink having been spilled about it, I doubt anyone has managed to do much better than this.

Nonetheless, here is how a professional philosopher, Susan Haack—in her seminal paper, "Six Signs of Scientism" (2012)—defines the object of our analysis: "What I meant by 'scientism' [is] a kind of over-enthusiastic and un-critically deferential attitude towards science, an inability to see or an unwillingness to acknowledge its fallibility, its limitations, and its potential dangers" (2012, 76). She goes on to explain that when the term *scientism* originated in the nineteenth century, it had a neutral connotation, simply meaning "the habit and mode of expression of a man of science" (ibid.). The more common negative connotation was acquired during the early decades of the twentieth century, apparently initially as a criticism of the overly optimistic claims of behaviorist psychologists regarding what a scientific approach to the study of human behavior could accomplish.

Nowadays, *scientism* can be used in three distinct ways, none of which is neutral. First, there is what I will call the *reasonable negative connotation*, along the lines of both definitions given above. This, for instance, is the use deployed by the philosopher Tom Sorell in his *Scientism: Philosophy and the Infatuation with Science* (1994; see also Sorell, this volume). Second, there is also on offer an increasingly pernicious *unreasonable negative connotation*, a simplistic use of *scientism* as an epithet to be hurled at ideas one doesn't like, in lieu of an argument (Boudry, this volume). For instance, the intelligent design (i.e., creationist) think tank Discovery Institute, together with the arch-conservative Heritage Foundation, hosted in 2015 a public forum entitled "Scientism in the Age of Obama," which was little more than a piece of political-religious propaganda with little or no intellectual content.[2] Third, there is a more recently emerging *positive connotation*, where some scientists and philosophers are attempting to "take back," so to speak, the term and argue something close to a negation of the two definitions given above, that is, that there is *no such thing* as an exaggerated trust in the efficacy of the methods of natural science applied to all areas of investigation. A good example of this is the philosopher Alex Rosenberg, in his *Atheist's Guide to Reality* (2011; see also Rosenberg, this volume). In the following, I shall be sympathetic (in a qualified manner) to the first approach, ignore as intellectually bankrupt the second one, and be critical (again, in a qualified manner) of the third one.

A Minimal Gallery of Scientistic Rogues

Before I get to a more detailed analysis of Haack's useful classification of "signs of scientism" (and flesh out my few but crucial disagreements with such analysis), let me give a few instances of what I consider to be obvious examples of scientism, just to help stake out my territory. One is *The Moral Landscape*, a book for the general public by author Sam Harris, in which he argues that science can *determine* human values, with no help from moral philosophy, which he dismisses in an endnote as "directly increas[ing] the amount of boredom in the universe" (2010, 211). This sort of thing has become a small cottage industry, with Harris himself later on publishing a book on free will in which he also advances the notion that science is all one needs to deal with the issue (and for which he has been thoroughly trashed by fellow New Atheist Daniel Dennett).[3] Author Michael Shermer contributed to the genre with his *The Moral Arc* (2015), which argues along lines that are difficult to distinguish from Harris's own. I have dealt in depth with both Harris's and Shermer's approaches to a science of morality,[4] but briefly, the reason I find them scientistic in nature is because of their intellectual totalitarianism: they do not seem to suggest that science can *aid* moral discourse (which would be reasonable, and also trivially true and thus not worth writing a book about), but rather that the scientific approach is the *only* one that can profitably do so.

A second class of pretty clear examples of scientism, in my book at the least, is represented by the dismissal—without argument—of humanistic disciplines by prominent scientists, often though not always physicists. Recent culprits are Stephen Hawking, Lawrence Krauss, and Neil deGrasse Tyson, among others. I will get to some of the details about their pronouncements near the end of this essay, and the interested reader is referred to my previous writings concerning some of these same episodes.[5]

A third class of scientistic examples comes from what has now become yet another cottage industry of articles and books that purport to provide *the* answer to a number of complex questions based on fMRI scans (or similar techniques) of the brain. This approach has been debated and debunked a number of times (Fine 2010; Satel and Lilienfeld 2013), but it persists in the media, and is in fact more popular than ever. Just like the case of the relationship between moral philosophy and science, the issue isn't that neuroimaging doesn't tell us *anything* of value about, say, gender differences, free will (again!), or a number of other interesting questions. The issue is that what neuroimaging tells us is always going to be only part of a complex story, and often not a particularly compelling part at that. It may be interesting, for instance, to pinpoint the neural correlates of moral decision making in humans (e.g. Greene and Haidt 2002), but to say that (indirectly estimated) activity in neuroanatomical structure X or Y *correlates* (statistically) with this or that type of moral thinking doesn't tell us even close to what we want to know about why people engage in certain types of moral thinking (let alone whether they are doing it *right*), which is presumably the result of genetic, developmental, environmental, and social factors, inextricably interlaced with one another. Indeed, if superficially understood, such research may even *harm* our understanding of whatever social-psychological phenomenon we are studying, by providing the illusion that we are nearing *the* crucial locus of explanation for it, while we are doing no such thing.

My brief gallery of rogues could continue with a number of other examples, but the main theme should by now be clear: I think scientistic thinking often originates from scientists who overstep epistemic boundaries (which I think of as real, albeit fuzzy, contra Thalos, this volume), if not of their discipline, certainly of their own expertise, and often of both. Again, this should not be mistaken for advocacy of a simplistically sharp demarcation between the humanities and the sciences, or philosophy of science and science. There are lots of reasons why the humanities and the sciences ought to cross-pollinate more often than they do in today's constraining academic world. But cross-pollination here is a two-way street, not a matter of intellectual imperialism on behalf of the natural sciences (*pace* Rosenberg, this volume). And it ought to be done thoughtfully and on the basis of actual knowledge, not starting from cardboard descriptions of a field of which one is (willfully) ignorant. Now that we have my version of the big picture more clearly at hand, let me discuss—as well as disagree with, when necessary—Haack's "six signs" of a scientistic attitude.

Haack's Signs of Scientism

These are the six attitudes that, in Haack's view, are unmistakable signals of scientism:

1. Using the words "science," "scientific," "scientifically," "scientist," etc., honorifically, as generic terms of epistemic praise.
2. Adopting the manners, the trappings, the technical terminology, etc., of the sciences, irrespective of their real usefulness.

3. A preoccupation with demarcation, i.e., with drawing a sharp line between genuine science, the real thing, and "pseudo-scientific" imposters.

4. A corresponding preoccupation with identifying the "scientific method," presumed to explain how the sciences have been so successful.

5. Looking to the sciences for answers to questions beyond their scope.

6. Denying or denigrating the legitimacy or the worth of other kinds of inquiry besides the scientific, or the value of human activities other than inquiry, such as poetry or art. (Haack 2012, 77–78)

There is no question that point 1 above does characterize what I, too, think are scientistic attitudes: to say that something is "scientific" becomes a substitute for actually giving thought to the matter. Examples range from popular advertisements ("9 out of 10 dentists recommend brand X"[6]) to serious misapplications of science, as in the well-documented case of eugenics (Bashford and Levine 2010).[7]

There are a number of examples of point 2 ready at hand, my current favorite perhaps being a much (and deservedly) criticized paper by Barbara Fredrickson and Marcial Losada, who argue—"scientific" data in hand—that the ratio of positive to negative emotions necessary for human flourishing is exactly 2.9013 to 1 (the current version of the paper has the estimate rounded off at 2.9) (Frederickson and Losada 2005, 683). Setting aside any (very pertinent) philosophical (and certainly not "scientific") discussion of what counts as human flourishing and why, Brown, Sokal, and Friedman critically analyze the highly impactful paper by Fredrickson and Losada, concluding:

We find no theoretical or empirical justification for the use of differential equations drawn from fluid dynamics, a subfield of physics, to describe changes in human emotions over time; furthermore, we demonstrate that the purported application of these equations contains numerous fundamental conceptual and mathematical errors. The lack of relevance of these equations and their incorrect application lead us to conclude that Fredrickson and Losada's claim to have demonstrated the existence of a critical minimum positivity ratio of 2.9013 is entirely unfounded. More generally, we urge future researchers to exercise caution in the use of advanced mathematical tools such as nonlinear dynamics and in particular to verify that the elementary conditions for their valid application have been met. (Brown, Sokal, and Friedman 2013, 801)

The fact that Fredrickson and Losada's was a rather obvious example of nonsense on stilts has not stopped their colleagues from citing the paper a whopping 322 times as of April 25, 2013, or from referring to such work as a "huge discovery." How was this possible? Because the authors deployed sophisticated mathematical tools to back up their bogus conclusions, and as we all know, mathematically informed science is *really* good science, so the claims *must* be reliable.

Haack's third criterion is the first one with which I will take issue. I do not know what personal experience might have solidified her view, but my own extensive dealings with a number of scientistically oriented authors have actually unearthed precisely the opposite of an obsession with demarcation (see also Thalos, this volume). Even superficially glancing at, for instance, the biologist Jerry Coyne's blog,[8] or skimming the above-mentioned book on science and morality by Harris (2010), will reveal that if anything, the scientistic strategy is one of expansionism, not demarcation. For the scientistic mind, *everything* worth inquiring on must be amenable to scientific analysis, by which it is meant not the specific type of systematized enterprise normally associated with professional science but rather *any* empirically based activity at all. Coyne's famous example is that of plumbing-qua-hypothesis-testing-hence-science, where he correctly observes that plumbing uses (roughly) the same analytical skills as scientific inquiry, yet bizarrely deduces from it, not that the two have something more basic—human reason—in common, but rather that there is no substantive difference between science and plumbing. I think this strategy is hopelessly misguided, since the corresponding expansion of the term *science* leads of course to its losing any meaning (recall that for some scientistically oriented people, even mathematics is, "ultimately, empirically based"),[9] and in fact it represents a throwback to the days of logical positivism/empiricism (Ayer [1936] 1956). That said, I do not endorse a strict demarcation between science and nonscience, or science and pseudoscience, either. While this was a research program popular in philosophy of science since Popper started it, we have seen that Laudan's (1983) classic paper put the brakes on it in its original form. As I mentioned above, there now seems to be a consensus among philosophers of science that demarcation is a worthwhile project, but that the outcome is going to be a series of inherently fuzzy boundaries based on something akin to Wittgenstein's family resemblance approach to the articulation of complex concepts (Pigliucci and Boudry 2013).

On her fourth sign of scientism, Haack is only partially on target. Despite the just-mentioned tendency of scientistically inclined authors to expand the domain of science beyond recognition, they still (must?) insist that there is a recognizable scientific methodology clearly superior to any other way of acquiring knowledge or understanding. But there's serious tension here for scientistic writers: on the one hand, they seem to want to make science and its methods coextensive with reason and empirical evidence (see point 3 above). On the other hand, there is a surprising amount of talk about "the" scientific method in science circles, for instance among theoretical physicists at a recent meeting on string theory I attended in Munich.[10] When queried, many scientists say they mean by that phrase something along the lines of the hypothetico-deductive approach, which squares well with widespread (and rather uncritical) acceptance of Popper and the idea of falsification. This is especially odd, as philosophy of science has moved well beyond logical empiricism and falsificationism by considering a panoply of new approaches, the foremost possibly being one version or another of Bayesianism (Bovens and Hartmann 2004). Then again, I suppose we can't expect scientists and scientistic authors to keep up with developments (dare I say progress?) in philosophy of science, despite the obvious relevance of that field to the matter at hand.

I entirely agree with Haack's fifth sign, having to do with a tendency to deploy science to address questions outside its proper domain of application. Indeed, I think I could go further and claim that for scientistically inclined thinkers there is no such thing as the proper domain of science, since nothing is outside its purview (see points 3 and 4 above). Defenders of scientism maintain that there simply are no examples of questions for which science does not or cannot in principle provide a full answer. And yet, it is easy enough to come up with a long list of these allegedly nonexisting examples. Consider, for instance, table 10.1, which is limited to philosophy (a much larger list could be produced if we were to expand the concern to the humanities more broadly).

Please note that my claim is *not* that the questions in this table cannot benefit from some sort of input from the sciences. Some do, for others it remains to be seen, and at any rate it would be foolish to impose limits on such input a priori. But a quick perusal of the table should make it painfully clear that science has little to do with a number of the entries, and has only partial input on the remainder—unless we fall back on the sleight of hand I mentioned above and redefine science in a way that is coextensive with reason

Table 10.1.

Area of philosophical inquiry	Examples of questions that cannot be settled by science
Metaphysics	What is a cause? How should we think about personal identity?
Logic	Is modus ponens a type of valid inference? What is the relationship between logic and mathematics?
Epistemology	Is knowledge justified true belief? Is an internalist or externalist perspective a better account of justification?
Ethics	Is abortion permissible once the fetus feels pain? Are there things that ought not to be for sale?
Aesthetics	Is there a principled difference between Mill's low and high pleasures? Is aesthetic judgment descriptive or prescriptive?
Philosophy of science	What logical role does genetic drift play in evolutionary theory? Is nonempirical confirmation a valid concept when it comes to theory assessment?
Philosophy of mathematics	What is the ontological status of numbers and other mathematical objects? Why is mathematics so surprisingly effective at aiding science?

itself, which would be not only historically and factually grossly inaccurate but ultimately meaningless. Or we could—as many scientistic writers are wont to do—deploy the trump card of Hume's famous dictum in the Enquiry (1748): "If we take in our hand any volume; of divinity or school metaphysics, for instance; let us ask, Does it contain any abstract reasoning concerning quantity or number? No. Does it contain any experimental reasoning concerning matter of fact and existence? No. Commit it then to the flames: for it can contain nothing but sophistry and illusion." Needless to say, however, neither Hume's own *Enquiry* nor pretty much anything written about scientism (both pro and con) would survive Hume's fork, so we should put it to rest once and for all. Indeed, I would go so far as to challenge my scientistically inclined colleagues who contributed to this volume to show me a single instance of systematic observation or experiment (i.e., an example of science) throughout this collection of essays. The contributions Maarten and I collected here are so inherently *philosophical* in nature that they stand as a self-evident refutation that science is our only path to knowledge and understanding.

Finally, we are left with considering Haack's sixth sign of scientism: the tendency to deny or denigrate the legitimacy of nonscientific activities, especially the humanities. Here we have an embarrassment of riches to draw from, in addition to those already briefly mentioned in my "rogues' gallery" above:

- "Philosophy is dead. Philosophy has not kept up with modern developments in science, particularly physics," says the eminent theoretical physicist Stephen Hawking (and Mlodinow 2010), right at the beginning of a book best characterized as a popular treatise on the philosophy of cosmology.

- Here is what the science popularizer Neil deGrasse Tyson said when he appeared on the Nerdist show back in 2014:[11] "My concern here is that the philosophers believe they are actually asking deep questions about nature. And to the scientist it's, what are you doing? Why are you concerning yourself with the meaning of meaning? . . . The scientist knows when the question 'What is the sound of one hand clapping?' is a pointless delay in our progress."

- The cosmologist Lawrence Krauss, in an interview with the *Atlantic*:[12] "Every time there's a leap in physics, it encroaches on these areas that philosophers have carefully sequestered away to themselves, and so then you have this natural resentment on the part of philosophers. . . . Philosophy is a field that, unfortunately, reminds me of that old Woody Allen joke, 'Those that can't do, teach, and those that can't teach, teach gym.' And the worst part of philosophy is the philosophy of science; the only people, as far as I can tell, that read work by philosophers of science are other philosophers of science."

- Steven Weinberg, Nobel physicist: "Philosophy of science, at its best seems to me a pleasing gloss on the history and discoveries of science . . . some of [philosophy is] written in a jargon so impenetrable that I can only think that it aimed at impressing those who confound obscurity with profundity. . . . The insights of philosophers have occasionally benefited physicists, but generally in a negative fashion—by protecting them from the preconceptions of other philosophers." (1994, 167)

Many, many more examples of this kind can easily be found in the literature. We could spend a significant amount of time rebutting both the specific points and the general outlook whence they spring (e.g. Pigliucci 2008), but this is beside the issue as far as this chapter is concerned. I hope that it is clear that the comments sampled above are the result of ignorance of the subject matter (none of these people have likely read much, if any, professional philosophy) and an astounding degree of anti-intellectualism, which is particularly disappointing when it comes from scientific luminaries or prominent science popularizers, all of whom ought to know better and have a duty to communicate responsibly with their readers.

Scientism and the Nature of Science

After years of having discussed the matter with both supporters and critics of scientism, I think it all really comes down to a single issue: how shall we think of science? The tug-of-war between the two factions is the result of scientistic writers attempting to expand our concept of science as much as possible, ideally making it coextensive with empirical evidence of any kind, or even with reasoning itself (it should go without saying that the latter is inherently broader than the former, as it includes, for instance, mathematics and logic). By that definition, then, literally everything we do, including logic, mathematics, and plumbing, is science. Pretty much the only nonscience activities left are religion and mysticism, which are usually treated with contempt by scientistic authors. (This does leave the arts, for instance, in a limbo, but even here a plethora of attempts at "explaining" artistic activity and aesthetic judgment have deployed either neuroscience or evolutionary biology, or both. While these are all interesting projects in their own right, they are also very partial—since they usually downplay or discard the historical-cultural milieu in which art is produced—and quite beside the point as far as artists and the art public are concerned.) On the other side, critics of scientism try to define science as narrowly as possible, pretty much to include only the academic disciplines of natural and possibly social sciences as understood today.

The whole thing, therefore, is an exercise in demarcation, and as we have seen, demarcation has suffered a loss of reputation in philosophy ever since Quine (1951) and his famous attack on "two dogmas" of empiricism, and then later on with Laudan (1983) and his rejection of the science-pseudoscience divide. And yet, it seems obvious that those distinctions are important, both conceptually and pragmatically, and that we can (indeed ought to) acknowledge the lack of sharp lines to be drawn anywhere without having thereby to embark on a slippery slope to the conclusion that no interesting distinction at all can be made.

Quine was arguably right that no clear-cut criterion separates analytic from synthetic truths; and yet, most of us (indeed, apparently most professional philosophers: Bourget and Chalmers 2013) have no trouble assigning "all bachelors are unmarried men" to the first one and "giraffes are vegetarian" to the second one, without any peril of confusion whatsoever (I think Quine was far closer to the mark with his second "dogma," reductionism). Similarly, Laudan was correct that no small set of necessary and jointly sufficient conditions will ever allow us to demarcate science or pseudoscience from other kinds of activities—and yet, most of us have no difficulty at all telling the difference between "scientific" creationism and evolutionary science.

The examples just mentioned should make it clear that demarcation is possible, useful, and here to stay (*pace* Thalos, this volume), although in all interesting cases it will take the form of a cluster concept, Wittgenstein-style. When applied to science, though, an additional criterion needs to be acknowledged: what it means to do "science" (and, therefore, what counts as nonscience) changes over time. Was Aristotle doing "science" when he was collecting and observing shells on the island of Lesbo? In a sense, but what he was doing was also pretty far from what modern biologists do, and not just in terms of technical advances. Was Ptolemy doing astronomy when he systematized a long-standing theory about the structure of the solar system? Again, sort of. But not in the way Copernicus and Galileo were doing it, and certainly nothing like what modern astronomers do.

It seems to me that by far the best way to think of (*define* is too restrictive) science is along the lines proposed by Helen Longino (1990; 2001). Science is an inherently social activity, dynamically circumscribed by its methods, its subject matters, its social customs (including peer review, grants, etc.), and its institutional role (inside and outside the academy, governmental agencies, the private sector). *That* is why it makes little sense to say that plumbing is science, or even that mathematics is science. Science is what scientists do, and scientists nowadays have a fairly clearly defined role, tools, and modus operandi, all of which easily distinguish them from plumbers and mathematicians, not to mention philosophers, historians, literary critics, and artists.

None of the above, of course, means that science cannot meaningfully contribute to nonscientific activities. It is surely interesting, for instance, to know that certain areas of the brain are stimulated by aesthetic judgment, or that human beings tend to find the contemplation of symmetrical features and the identification of complex patterns inherently pleasurable, plausibly for evolutionary reasons (but *only* plausibly: Kaplan 2002). But that sort of scientific knowledge doesn't even begin to exhaust both art as a human activity and its reception and significance in human societies, at the least not until we also bring in the humanities, cultural history, and the social sciences— but that would be "understanding" art just in the way it is, in fact, currently understood.

A colleague recently gave a guest lecture in a class I have been teaching on epistemology across the curriculum, and he brought up yet another example of the shortsightedness of scientism. His research falls within the domain of the historical and social sciences, particularly focusing on Western colonialism during the sixteenth through the nineteenth century. Part of that type of scholarship relies on historical data, part on "scientific" approaches, for instance when it comes to quantifying different parameters describing the history and impact of colonialism. But part of it is founded on reading the first-person accounts, both nonfictional and even fictional, of people who actually experienced colonialism at first hand. It is *not* the case that first-person experience trumps everything else, because, of course, it provides a very narrow and biased perspective. But it *is* the case that without it we would have a much harder time understanding what colonialism meant and how it worked. My colleague's scholarship, therefore, lies at the interface between social science, history (traditionally a humanistic discipline), and literature. The science part makes a crucial contribution to his work, but he simply cannot do without the other two, no matter how loudly supporters of scientistic imperialism scream.

In the end, I am reminded of a quip by Alan Sokal, author of the famous hoax that helped to debunk the most extreme postmodernist notions during the science wars of the nineties: "When one analyzes [postmodernist writings on science], one often finds radical-sounding assertions whose meaning is ambiguous and that can be given two alternative readings: one as interesting, radical, and grossly false; the other boring and trivially true" (1998, 13). Now simply substitute [scientistic] for [postmodernist] and [on knowledge] for [on science] and you have a pretty good summary of the situation at hand: yes, science is crucial in order to further our understanding of the world in which we live, including human culture. But no, it isn't the beginning and the end of such understanding.

······

Notes to Chapter Ten

1. Although Popper's work cited here is the first American edition of his Conjectures and Refutations, which appeared in 1963, his ideas on this subject date back at the least from *Logik der Forschung* (Vienna: Julius Springer Verlag), which was published in 1935 and translated as *The Logic of Scientific Discovery* (London: Hutchinson) in 1959.

2. Heritage Foundation, "Scientism in the Age of Obama," accessed January 25, 2017. http://goo.gl/S7Zpwv.

3. Daniel Dennett, "Reflections on 'Free Will,'" January 24, 2014, Naturalism.org, http://goo.gl/mALMlN.

4. On Harris: Massimo Pigliucci, "Science and the Is/Ought Problem," *eSkeptic*, February 2, 2011, http://goo.gl/ZE6g07; on Shermer: "Dialogue: What Is the Role of Science in Morality?," NorthEast Conference on Science and Skepticism; *Rationally Speaking* podcast featuring Massimo Pigliucci and Michael Shermer, hosted by Julia Galef; YouTube video, posted May 1, 2013, 34:54; https://goo.gl/9azTn9.

5. Massimo Pigliucci, "Lawrence Krauss: Another Physicist with an Anti-philosophy Complex," *Rationally Speaking* (blog), April 25, 2013, http://goo.gl/bZtGSh; Massimo Pigliucci, "Neil deGrasse Tyson and the Value of Philosophy," ScientiaSalon.org, May 12, 2014, https://goo.gl/ctmE9K.

6. See, for instance, the website for Sensodyne, "No. 1 Dentist Recommended Toothpaste for Sensitive Teeth," accessed February 16, 2017, http://goo.gl/GgPBhZ. Similar examples are not at all hard to find.

7. See a disturbing "Eugenic certificate" dating to 1924 on the Disability History Museum website, accessed January 25, 2017: http://goo.gl/QuNaZn.

8. Jerry Coyne, *Why Evolution Is True* (blog), accessed January 25, 2017, https://goo.gl/1yVDJO.

9. "The Limits of Science," discussion by Daniel Dennett, Lawrence Krauss, and Massimo Pigliucci on *Het Denkgelag*, YouTube video, posted November 5, 2013, 1:56:19; https://goo.gl/AUUObK.

10. See Massimo Pigliucci, "Why Trust a Theory?—part I," *Footnotes to Plato* (blog), December 8, 2015, https://goo.gl/HmCtYR; "Why Trust a Theory?—part II," December 9, 2015, https://goo.gl/Xa4L2q; and "Why Trust a Theory?—part III," December 10, 2015, https://goo.gl/Cm8ysn.

11. "Neil deGrasse Tyson Returns Again," Nerdist.com podcast audio posted by Katie Levine on March 7, 2014, http://goo.gl/A7DDjN (the relevant bit starts at 20'19").

12. Quoted in Ross Anderson, "Has Physics Made Philosophy and Religion Obsolete?," *Atlantic*, April 23, 2012, http://goo.gl/pOjrSJ.

Literature Cited

Ariew, R. 1984. "The Duhem Thesis." *British Journal for the Philosophy of Science* 35: 313–25.

Ayer, A. J. (1936) 1956. *Language, Truth, and Logic*. Mineola, NY: Courier Corporation.

Bashford, A., and P. Levine. 2010. *The Oxford Handbook of the History of Eugenics*. Oxford: Oxford University Press.

Bourget, D., and D. J. Chalmers. 2013. "What Do Philosophers Believe?" *Philosophical Studies* 3: 1–36.

Bovens, L., and S. Hartmann. 2004. *Bayesian Epistemology*. Oxford: Oxford University Press.

Brown, N. J. L., A. D. Sokal, and H. L. Friedman. 2013. "The Complex Dynamics of Wishful Thinking: The Critical Positivity Ratio." *American Psychologist* 68: 801–13.

Fine, C. 2010. *Delusions of Gender: How Our Minds, Society, and Neurosexism Create Difference*. New York: W. W. Norton.

Fredrickson, B. L., and M. F. Losada. 2005. "Positive Affect and the Complex Dynamics of Human Flourishing." *American Psychologist* 60: 678–86.

Greene, J., and J. Haidt. 2002. "How (and Where) Does Moral Judgment Work?" *Trends in Cognitive Sciences* 6: 517–23.

Haack, S. 2012. "Six Signs of Scientism." *Logos and Episteme* 3 (1): 75–95.

Harris, S. 2010. *The Moral Landscape: How Science Can Determine Human Values*. New York: Free Press.

Hawking, S., and L. Mlodinow. 2010. *The Grand Design*. New York: Bantam.

Hume, D. 1748. *An Enquiry concerning Human Understanding*, http://goo.gl/Gd4YPz (accessed January 25, 2017).

Kaplan, J. M. 2002. "Historical Evidence and Human Adaptation." *Philosophy of Science* 69: S294–S304.

Laudan, L. 1983. "The Demise of the Demarcation Problem." In *Physics, Philosophy and Psychoanalysis*, edited by R. S. Cohen and L. Laudan, 111–27. Dordrecht, Netherlands: Reidel.

Longino, H. 1990. *Science as Social Knowledge: Values and Objectivity in Scientific Inquiry*. Princeton, NJ: Princeton University Press.

———. 2001. *The Fate of Knowledge*. Princeton, NJ: Princeton University Press.

Pigliucci, M. 2008. "The Borderlands between Science and Philosophy: An Introduction." *Quarterly Review of Biology* 83: 7–15.

———. 2013. "A (Belated) Response to Laudan." In *Philosophy of Pseudoscience: Reconsidering the Demarcation Problem*, edited by M. Pigliucci and M. Boudry, 9–28. Chicago: University of Chicago Press.

Pigliucci, M., and M. Boudry, eds. 2013. *Philosophy of Pseudoscience: Reconsidering the Demarcation Problem*. Chicago: University of Chicago Press.

Popper, K. 1963. *Conjectures and Refutations: The Growth of Scientific Knowledge*. New York: Basic Books.

———. 1978. "Natural Selection and the Emergence of Mind." *Dialectica* 32: 339–55.

Quine, W. V. O. 1951. "Two Dogmas of Empiricism." *Philosophical Review* 60: 20–43.

Rosenberg, A. 2011. *The Atheist's Guide to Reality: Enjoying Life without Illusions*. New York: W. W. Norton.

Satel, S., and S. O. Lilienfeld. 2013. *Brainwashed: The Seductive Appeal of Mindless Neuroscience*. New York: Basic Books.

Shermer, M. 2015. *The Moral Arc: How Science Makes Us Better People*. New York: Henry Holt.

Sokal, A. 1998. "What the Social Text Affair Does and Does Not Prove." In *A House Built on Sand: Exposing Postmodernist Myths about Science*, edited by N. Koertge, 9–22. Oxford: Oxford University Press.

Sorell, T. 1994. *Scientism: Philosophy and the Infatuation with Science*. London: Routledge.

Thornton, S. 2013. "Karl Popper." *The Stanford Encyclopedia of Philosophy* (Winter 2016 ed., with substantive content change), edited by Edward N. Zalta, http://plato .stanford.edu/entries/popper/.

Weinberg, S. 1994. "Against Philosophy." In *Dreams of a Final Theory: The Scientist's Search for the Ultimate Laws of Nature*, 166–90. New York: Vintage.

11

Strong Scientism and Its Research Agenda

ALEX ROSENBERG

Introduction

For scientism most of metaphysics is easy. Almost all of it can pretty much be read off of science: the physical facts fix all the facts. There is no meaning to the universe, life and lives have no purpose, the mind is the brain, there is no free will or soul or even an enduring numerically identical self for that matter. Scientism makes equally short work of the normative realm. Metaethics is unproblematic: no ethical theory is grounded more firmly than any other, and whichever one humans adopt is a matter of natural and cultural Darwinian selection. Substantive ethics is of course a non-starter for scientism. This extremely tendentious set of claims apparently leaves many hostages to fortune. All of these claims are controversial to say the least and most opponents of scientism have what they consider devastating counter-arguments. This is not the place either to fully expound the arguments for the claims of strong scientism (but see Rosenberg, 2011 for a book-length exposition), or for a defense against well-known objections to scientism that stem from equally or even more tendentious premises than the ones scientism starts from. The aim of this chapter is to identify those matters, mainly epistemological, that should give those of us who embrace scientism real pause: the agenda of challenges scientism has to deal with for its own adherents to increase their confidence in its adequacy.

You might suppose that scientism can make equally quick work of epistemology. Surely all we need is the scientific method. Now, establishing exactly what that method should be is by no means easy. After all, different sciences perforce employ methods suited to their own domains and not those of others. It's perfectly clear that for finite epistemic agents like us, for example, the methods of physics—even if there is one set of methods that are employed in physics—are not going to secure knowledge in almost any area beyond physical chemistry. They won't tell us much of what we want to know about the biological domain. It's true that the laws which physics discovers by the use of its methods do prescribe Darwinian processes as the only means by which adaptation can emerge anywhere in the universe. (For details, see Rosenberg 2006, chap. 6; Rosenberg 2011, chaps. 4 and 4.) But it's doubtful that methods and data from physics would enable creatures like us to derive Darwin's theory *ab initio*, as the physicist understands it.

There may not be a general scientific method that both informatively and correctly describes the methods of all the sciences. The injunction to operate inductively may be correct, but it is insufficiently informative. Narrower methodological principles, on the other hand, such as Ockham's razor or inference to the best explanation or model building, and so on, must be applied with discernment, not to say artfulness. Science will generate no algorithm for producing the right rules or for when to apply them, except possibly post hoc, when it is too late.

Is it enough for scientism simply to pass off the obligation to offer an epistemology to science? Can scientism simply say that the sciences will ground the epistemology we should use to build our metaphysics—science? We can argue that each domain will build its proprietary epistemology, driven by the nature of its research domain. Each will be subject to the requirement of consistency with the epistemology of more fundamental science—and we know well the order of fundamentality. And each will help itself to breakthrough methods and instruments offered and epistemically certified by advances in the more fundamental disciplines. That's all we need to say, right? Alas, no. This is largely because challenges to scientism are almost always epistemic. How do you know scientism is true? How can scientism account for our knowledge of this or that domain of knowledge, usually mathematics and logic, sometimes normative ethics, history or human affairs generally, occasionally revealed religion? In the latter cases, scientism doesn't have much trouble writing off the purported domains as ones in which there is no knowledge, only literary and creative products with a variety of political, emotional, or aesthetic functions. But mathematics and logic cannot be written off so blandly. A defense of scientism against challenges from these domains requires more than promissory notes about what epistemology will be like when science is complete.

In what follows, I sketch the demands on and constraints facing an epistemology for scientism, and then try to wield them into tentative suggestions for how it can meet its most serious challenges.

Epistemology for the Scientistic

Scientism's epistemology has to be *naturalistic*. *Naturalistic epistemology* is a term of art from Quine (1969), if not from Dewey (1938), that has been attractively developed in the hands of philosophers such as Hilary Kornblith (2015) for a score of years or more. As Quine pointed out, for the naturalist there can be no first philosophy prior to science that somehow certifies its results as knowledge. We perforce reject the notion, inherited from Descartes and so freighted by Kant's *Critique of Pure Reason*, that without foundations in pure reason, science's results could not claim to be knowledge. Naturalistic epistemology holds that knowing is a cognitive state with causes and effects, and more important, that when X knows that P, there is a causal connection between (the facts that make) P (true) and X's knowledge of P.

The task of epistemology on this view is to determine which causal chains reliably provide true beliefs, and which do not. Once these epistemically reliable causal chains are identified, the epistemologist is in a position to certify some beliefs as justified and true and so (*modulo* Gettier counterexamples) as knowledge. Scientism is entirely sympathetic to this program. It adds what is perhaps implicit in naturalistic epistemology, that the only way to identify these reliable causal processes is by the means of science.

But then how does naturalism—scientistic or otherwise—respond to the circularity charge: without a prior conception of knowledge and a criterion for when it is achieved, how can science claim for itself the status of knowledge—justified true belief? It can't appeal to itself.

Some may say that science doesn't have to. Science is corrigible, and many of its claims will be tentative; most of its claims at the frontiers of knowledge will not be deemed knowledge, especially by their promulgators. Some scientists may even be global skeptics (e.g. van Fraassen 2002), doubting that we can really be sure of anything in science. Consider how so secure an edifice as Newtonian mechanics was toppled in a few short decades. Think about the incompatibility of our two best physical theories—quantum mechanics and general relativity. Good reasons to be modest in our pretensions to knowledge. Some serious scientists (e.g. Susskind 2014) deny science can, does, or should provide knowledge, rather than just practically successful instruments for organizing "experience," whatever that exactly is. It provides justified belief, where justification is a matter of practical success in attaining our technological aims.

Practically justified belief may be enough for scientists, and even more congenial for engineers. Even some philosophers might take "justified" to be a matter of some properly qualified long term improving predictive success. And surely even those scientists who eschew "knowledge" because they can't establish the truth of their most well-justified findings and theories are willing to certify them as justified, or at least more justified than their predecessors.

The epistemology of scientists who adopt this view, and philosophers who agree with them, is of course not naturalism. It's pragmatism, a view that I for one think incompatible with naturalism. In fact, I am inclined to call it Protagorianism—for its adoption of the maxim that "man is the measure of all things—of what there is, and what there is not." (Cf. Rorty 1981; Dewey 1938.) Pragmatism certifies as knowledge whatever works for us, and designates as truths what science, "in the limit and when complete," contains by way of theories and findings. This latter claim may be something pragmatism and scientism share. But scientism can't hold that either now or in the limit, scientific truth is whatever works for us.

The reason scientism can't be pragmatic in its epistemology starts with its metaphysics. Science tells us that we are components of the natural realm, indeed latecomers in the scheme of things that go back 13.7 billion years. The universe wasn't organized around our needs and abilities, and what works for us is just a set of contingent facts that could have been otherwise. Among the explananda of the sciences is the set of things—including beliefs (let me have this notion for the nonce)—that work for us. Once we have begun discovering things about the universe that work for us, science sets out to explain why they do so. It's clear that one explanation for why things work for us that we have to rule out as unilluminating, indeed question-begging, is that they work for us because they work for us. If something works for us, enables us to meet our needs and wants, then there has to be an explanation for why it does so, reflecting facts about us and the world that produce the needs and the means to satisfy them.

The explanation of why some methods work for us must be a causal explanation, broadly speaking. It must show what facts about reality brought it about that we discovered methods that produce predictively successful beliefs about reality. The explanation has to show that their working is not a coincidence, still less a miracle or accident. That means there have to be some facts, events, processes that operate in reality and which brought our pragmatic success about. This requirement immediately prohibits scientism from adopting the agnostic attitude of some scientists that all science can, does, or needs to provide is successful technological application. The demand that success be explained is a consequence of science's epistemology.

So our epistemology, our methodology for securing practically useful, predictively reliable beliefs about the world, has to be a set of rules underwritten by a causal explanation of why they work. Here's an example: consider the methodological rule that in pharmacological and other experiments on people, double-blind procedures enhance the reliability of experiments in securing knowledge. The explanation for why this rule works relies on such facts as the existence of a placebo effect on subjects and an "experimenter" (Rosenthal and Jacobson 1968) effect on scientists. These two effects together explain why the double-blind rule works in pharmacology, or at least they explain why using it is a necessary condition for securing reliable results. It's a simple example of the general character of methodological, that is, epistemic justification that scientism accepts: the norms of reliable information acquisition (call it knowledge for the moment) are those that reflect the causal processes operating in the universe. They are hypothetical imperatives whose effectiveness reflects the operation of regularities in their domains of application that can be explained, usually by considerations from more fundamental scientific theories. This will work all the way down to norms of research employed in fundamental physical research, where the empirical results are firmly established but the methodological rules lack a deeper explanation, as might be expected at the basement level of science.

The epistemology of scientism is given by the norms of scientific methodology, which in turn are justified as heuristic rules, hypothetical imperatives, by scientific theories that explain why they work. These theories are rarely if ever ones uncovered in the domains whose methods they underwrite. But what gives us confidence—epistemic warrant—that the explanations cum justification for the rules that these theories provide are true, or approximately true, or increasingly approximately true? Here there is a limited role for prediction. Why suppose that a putative explanation is an actual explanation or even part of the correct explanation or on the right track toward an explanation of anything, a methodological norm, regularity in phenomena, a model that works or even a theory? Scientism has to reject several alternative accounts of explanation because like a pragmatic epistemology, they ground epistemic warrant on satisfying human needs or wants. These are the erotetic, pragmatic, or unifications accounts of explanation (see Salmon 1989). These three accounts of explanation between them explain a great deal about what humans rightly or wrongly take to be explanations, but they reveal that too often what we accept is a matter of the circumstances in which individuals ask why questions, and reflect facts about human cognitive styles and computational limitations.

Scientism needs a standard of explanation that will be endorsed by agents of far greater cognitive powers than our current ones. How do we tell whether as an objective matter a putative explanation really explains, really identifies the causes of the explanantia? Or at least whether it provides a better, an improved explanation that supersedes a previously accepted one? Here predictive success plays a modest epistemic role: an explanation needs to have some predictive consequences, and a better explanation must eventually have some new predictive consequences, either improved in precision or in range of phenomena, either by itself or in conjunction with other theories and explanations, either sooner or later. The predictions that certify an explanation as good or better may be indirect, even practically inconsequential. But their role is indispensable. Contriving putative explanations that satisfy humans, that reduce the itch of curiosity, that exploit our preconceptions, is easy. The label *just so story* Gould and Lewontin (1979) employed to stigmatize adaptationist explanations appropriately attaches to explanations in all the domains of science. The meretricious appeal of just so stories held science back for a couple of thousand years (cf. Weinberg 2015).

The source of many of these stories, and the standards that urge that they stand to reason, is "common sense." Science tells scientism to be extremely suspicious of epistemic norms arising from this source or urged by introspective reason, even endorsed by it as indubitable. Not only is the track record for common sense poor, but cognitive neuroscience has shown that most of what conscious experience and reflection tell us about the world, and about rules for providing ourselves with knowledge of the world, is unreliable, especially in circumstances where quantitative inference is involved. (Cf. Loftus 2008; Purves et al. 2012; Weiskranz 2009.)

Of course, scientism's epistemology rules out many domains of human endeavor that others claim to be knowledge: much of interpretive social science and history, psychoanalytic and psychodynamic "theory," cultural studies and human affairs generally, and of course the literary humanities. In reflecting on why scientism cannot accord achievements in these domains epistemic standing, two things have to be borne in mind: first, that almost all the deliverances of disciplines in these domains have important (but not cognitive) psychological, social, and cultural effects, and there are great differences among them in the successes with which they fulfill these noncognitive functions. The second thing to bear in mind about them is that though these human productions don't constitute knowledge by the lights of scientism, they may well, and in some cases mostly, express true (but not explanatorily significant) claims about facts (*modulo* an important qualification to be made below).

Consider history, for example. Why must scientism reject its claims to knowledge? Well, to begin with, scientism rejects any causal or nomological or quantitative explanation/interpretive understanding distinction. If the commonsense "theory of mind" behind interpretive understanding of the sort that constitutes most history and biography cannot pass the weak predictive test, then when historical explanations actually do use it to identify the considerations causally relevant to their explananda, they are at most right by accident. But folk psychology—the theory of mind that animates these explanations—cannot pass reasonable tests of predictive improvement. It wasn't too bad as a tool for getting humankind from the bottom of the food chain on the African savanna to the top in a matter of a million years or so. But it is not much better than that, as cognitive science and neuroscience between them show. Scientism diagnoses the problems of folk psychological ("theory of mind") explanation as the result of a mistaken view that the brain represents content, and does so sententially, in neural circuits that express propositionally expressed thoughts. The assumption that it does so is not a harmless idealization but a significant cause of the predictive and explanatory failures of every theory that makes it. Add to this harmful assumption the fact that like all Darwinian processes, cultural change proceeds in a series of ever-shortening local equilibrium and ever-accelerating arms races, and the failure of the interpretive disciplines to convey real explanatory knowledge is the unavoidable conclusion from the epistemological position of scientism.

It's no part of scientism's epistemic obligations to explain why the interpretive disciplines persist, and why humans generally cannot surrender the conviction that they do provide significant increases in knowledge (though it is an obligation of science itself to do so). Answers to these questions require the resources of cognitive neuroscience. But scientism's admiration for the emotive force, entertainment value, aesthetic power of the great works of interpretive understanding is second to none. Solzhenitsyn's diagnosis in the *Gulag Archipelago* of why Soviet Communism produced its system of repression may or may not be factually correct, but its impact on readers, on public opinion, on the course of history, cannot be questioned. And none of these effects require that it constitute knowledge.

Scientism's Problem with Mathematics

Like every naturalistic epistemology, scientism has trouble with mathematics (and probably also logic). If we take mathematics at face value, its domain of numbers, sets, functions, lines, points, and so on comprises abstract objects, that is, unique particular things, like planet Earth, albeit ones that don't exist in space and time, have no parts, size, shape, mass, or other dimensions associated with concrete particular objects. This of course prohibits us from having any causal relationship with them. Accordingly, on a naturalistic epistemology we cannot have any knowledge of them, and of the relations among them.

This conclusion would not just be a fly in the ointment of all naturalistic epistemologies, including scientism's. It is, in the view of some, sufficient reason to reject scientism's epistemology altogether, and indeed all naturalistic epistemologies. For there seem to be powerful reasons to suppose that we have mathematical knowledge, that mathematical knowledge is the most secure knowledge that there is, that is, many of our mathematical beliefs are absolutely true and more securely justified than any other beliefs we have. If we can't have mathematical knowledge, we can't have any knowledge and are reduced to skepticism, or at least open to challenges from skeptics. Even if you are not yourself a mathematician, the one fact about your mathematical education that differs from every other discipline you studied is that in this domain you never had to unlearn anything you were taught. In every other domain or discipline, we begin with simplifications, approximations, and idealizations. In mathematics, from the beginning you learn propositions that never need correction, qualification, completion, and if your teacher is any good you learn compelling arguments for these propositions. Mathematics is the domain of knowledge par excellence.

Almost invariably, philosophically sophisticated opponents of scientism offer our knowledge of mathematical truths and knowledge of the existence of their truth makers as the most serious objections to its claims. The combination of their undeniable truth, their insulation from any form of causal inter-action with us, and their indispensability in science itself makes mathematics the problem from hell for scientism.

Of course, it's not just scientism, or more broadly, naturalism, that has problems with mathematics. Empiricism long had problems with *abstracta*, especially numbers. Hume celebrated Berkeley for his "wonderful discovery in the republic of letters," that what we think are abstract objects turn out to be just words—tokens in thought, sound, and inscription mistaken for them. Alas, the solution to the problem of mathematical knowledge was not so simple. Empiricism had to account for the certainty, necessity, and univer-sality of mathematical truths—the features that explain why you never have to unlearn any of it. Empiricists thought they had a solution to this problem when they claimed mathematical truth to be definitions, and logical conse-quences of definitions, where logical consequence is also a matter of defi-nition. It took until the 1930s for Gödel to explain why the "logistic" program for solving empiricism's problem about mathematical certainty could never succeed. Gödel himself, like other mathematicians, was driven to insist that numbers are abstract particulars, and that since we have knowledge of them, naturalism is false. This is a view shared all the way back to Plato, of course. (Cf. Shapiro 2000.)

Can rationalism solve these problems? Rationalism is the thesis that the mind has a priori knowledge, including knowledge of mathematical truths or at least truths from which mathematics can be erected, constructed, derived. It's easy to assert rationalism, but mere assertion doesn't actually effect the construction of mathematical knowledge. We must not confuse rationalism with theories that accord the mind, that is, the brain, innate ideas, and hold that among these are mathematical ones. Innateness is a psychological claim about the origin of ideas, concepts, thoughts, not an epistemological claim about their warrant or a metaphysical claim about the ontological status of their referents. Abstract objects can no more have effects on the mind before it acquires sensory awareness than they can after it begins to do so. Ab-stract objects and human thought are completely sealed off from each other. Whether our numerical concepts are caused by experience or emerge from hard-wired neural circuitry doesn't actually do any work in solving the prob-lem of mathematical knowledge.

Can rationalism solve the problem of mathematical knowledge by adopt-ing some dualist theory of mind and body according to which the mind is nonphysical, and therefore somehow has nonexperiential, noncausal access to mathematical objects and truths about them? If the mind is not physical but located somewhere inside the body, it is hard to see how it could actually do anything—it certainly can't do anything physical, such as change its state, including physically representing to itself psychologically novel mathematical truths. How a breathless, lengthless, depthless particular object can do anything at all is beyond our comprehension. So this version of rationalism doesn't seem to be a solution to the problem, just a matter of piling on more epistemological problems.

In short, if there are abstract mathematical objects and if we have knowledge of the relations among them, then every epistemology taught in the schools is in trouble. (In what follows, I will assume that the various attempts among contemporary metaphysicians to avoid taking quantification over mathematical objects seriously will not succeed. But cf. Yablo 2009; Chalmers 2009 for discussion of this sort of gambit.)

There is a *tu quoque* argument open to scientism in response to its critics. No one has a satisfactory epistemology of mathematics; no one has a satisfactory ontological account of the existence of numbers. So on this basis there is no reason to prefer other packages of metaphysics and epistemology to scientism's package. And then when we add the arguments for scientism outside this domain into the scales, the balance tips toward scientism.

Well, this may be a response to objections, but it is no solution to the problem. Insofar as scientism lays claim to being an adequate philosophy, it needs to deal with the problem of mathematical knowledge and mathematical objects. Of course, if the options facing scientism in doing so were attractive, they'd also be helpful to less radical doctrines such as naturalism. If some version of *nominalism* were true, as Hume and Berkeley hoped, scientism, naturalism, and empiricism could deny that there are numbers, holding instead that there were only numerals, including ones in our (physically concrete) thoughts, and that we can make sense of statements about numbers without quantifying over them, by translating them into statements about numerals. As for the grounds of our knowledge of mathematical truths, scientism, naturalism, and empiricism can adopt John Stuart Mill's view that all the truths we learn about the relationships among numbers are inductive generalizations from instances of these truths we meet with in experience. We could extend this theory to include, besides enumerative induction, inference to the best explanations of the inductively supported truths, and reasonably hope that this form of reasoning will give us all the higher math we want to adopt, all the way up to the four color theorem and the proof of Fermat's theorem.

Naturally, Mill's treatment of mathematical truths won't give us any of the truths that objectors demand we provide causal grounds to accept. For the mathematical truths they demand be certified as knowledge are necessary truths, and the demand is that besides their truth scientism (naturalism, and empiricism) certify that we know their necessary truth, and their necessary nontautological truth for that matter (remember Gödel).

As an exponent of scientism, I would readily settle for an epistemology that gives me all the truths of mathematics as well grounded without going to the lengths of being necessarily true. Scientism does not take sides on the nature of modality or even its indispensability in a scientistic metaphysics or logic.

One attractive approach to the problem of mathematical knowledge for the naturalistic epistemologies also happens to be one that accords especially well with scientism. This approach begins by denying the indispensability of mathematics to science. To this claim it adds the radical claim that mathematical statements are all false, since their truth requires the existence of numbers and other *abstracta*. However, among the falsehoods there are important, useful statements, statements that are, so to speak, true-in-the-story that scientifically useful mathematics tells. Fictionalism about mathematics accepts that "2 + 2 = 4" is, like "Holmes was a consulting detective," a fiction, a statement that is strictly false, but an important part of a calculating device that makes short work of otherwise insoluble scientific problems. (For an introduction to fictionalism about mathematics, see Balaguer 2011.)

Some proponents of fictionalism about mathematics have separated it from the arguments against the indispensability of mathematics to science. They accept that science really is committed to the existence of numbers and other abstract objects, and therefore it is committed to some falsehoods. (Cf. Kitcher 1984.) They assuage any discomfort about this conclusion by dividing the commitments of science into ones about concrete entities and relations—particles, field, and their relations—and ones about *abstracta* like numbers. The difference between these commitments is that the former are true (or at least approximately true) and the latter are false (though useful). This is not an option open to scientism, as it would require an epistemic basis other than science itself for our ability to distinguish the true commitments of science from the false ones.

If scientism is going to accept fictionalism about mathematics, it is also going to need to adopt the dispensability for science of its literal truth. For scientism, the price of the package of fictionalism and indispensability is too high.

What are the prospects of showing that numbers are dispensable and statements about them are fictitious? Hartry Field (1980) has made a start, showing that we can do Newtonian mechanics without quantifying over numbers. The trouble is it looks like Field's approach and others like it require that we quantify over sets. Can scientism and other naturalisms accept sets with equanimity? Perhaps some work of David Lewis (1993) can help.

Like others, Lewis holds that mathematics, or at least most of it, is reducible to set theory, that numbers can be "generated" from classes, and in particular that they can all be generated from the singleton class, with one member. Take any individual thing: there is, besides that thing, the singleton class of which it is a member. Given some basic axioms Lewis develops that have all the obviousness of the first of them (one class is part of another if the first is a subclass of the second), Lewis claims to be able to build up almost all mathematics from these minimal commitments to *abstracta*. And they are mighty minimal: we start with a perfectly concrete thing, a chair or table or knife or fork, and then recognize the class of things consisting of it and it alone. This singleton class is different from the individual item that is its only member, it is abstract but is not physically distinguishable from its member, and there seems no room for or reason why scientism, or naturalism, or empiricism, should have any qualms about its existence opening up floodgates to other, more suspect kinds of *abstracta*. All we need to accept is that in having causal access to the individual we also have causal access to its singleton class, and then to the class of the chair and its singleton class, and so on. Is this so hard? If Lewis's program is as good as his word, and the singleton class is the sole basic kind of *abstractum* we need to admit to in order to have our mathematics and eat our scientism, too, then there may be hope for an answer to the opponents' challenge from mathematics. At any rate, the problem for scientism, and for the other naturalisms, will be a clear and limited one: find a causal theory of our knowledge of singleton sets, given our complete knowledge of their members. And if no such causally underwritten knowledge is in the offing, we can always try treating the singleton set like Sherlock Holmes.

Scientism's Even More Serious Problem with Intentionality

Scientism has to be eliminativist about propositional attitudes. It has to deny that there are sentences represented in the neural circuitry, original intentionality in Searle's convenient label. It has to deny that there are thoughts *about* anything. It doesn't have to deny there are thoughts. It just has to deny that they are *about* something, with a "topic" on which they make "comments," in Dretske's nice nontechnical terms.

It would be nice if scientism didn't have to be eliminativist about mental content. It would save advocates of scientism a lot of hard work and remove the greatest impediment to its acceptance. But alas, it is not to be. Scientism has to deny that thought, brain states, the neural circuitry has intentional content because that's pretty much what neuroscience forces it to accept. The brain is nothing but neural circuitry, logic gates through which currents flow all the way from where the peripheral nervous system meets the world to the afferent processes that move the body and its parts. Now, there are several ways to see why scientism has no choice but to go eliminativist. First way: think about Watson, the IBM computer that beats human champions at *Jeopardy!* Combine the achievement of the Watson group at IBM with the insight of Searle's Chinese Room, and it becomes clear that Watson has only derived intentionality. There is no original intentionality of any amount in all those integrated circuits, just logic gates steering changes in charge distributions. All the rest is our interpretation (whatever that could be). It's we who accord Watson its derived intentionality. Now consider the brain of one of Watson's software engineers, one of the humans who confer derived intentionality on Watson's hard drive memory store. The scientist's brain is, like Watson's brain, a (much larger) set of integrated circuits, logic gates, steering changes in charge distributions. Where does its intentionality come from? Well, it can't come from one of the software engineer's colleagues, family members, or any other human who interacts with the engineer. Therein lies infinite regress. Dualists may think they can solve this problem, but I don't see how a nonspatial thing at a mathematical point in space can change its content, even if you assume it could have any. Dualism isn't an alternative for scientism.

Some will argue, like Searle (1980), and Horgan and Tienson (2002), that it's the engineer's consciousness that bears original intentionality, and that allows him or her to confer derived intentionality on Watson. But this won't do for scientism either (it shouldn't do for anyone): what's intrinsically originally intentional about tokens moving through consciousness? Their appearance of intentionality is just a matter of the pattern of succession and association of these tokens passing through consciousness. Think of the tokens passing through the conscious states of the newborn infant. What are they "about"? What is more, scientism has to eschew any confidence in the claims of introspection, not merely owing to what cognitive neuroscience has taught us in recent years about how wrong it is. The whole history of science is the history of the repeated and recursive reconstruction of introspective reason and common sense by the elimination and replacement of one false claim after the other, in spite of the fact that each are vouched safe by conscious experience.

A second deep pillar at the foundations of eliminativism is provided by the molecular biology of learning, memory. Eric Kandel won the Nobel Prize for showing how classically conditioned learning is realized in the neurology of the sea slug. Roughly, short-term learning is a matter of changes in the concentration of neurotransmitter molecules and sodium, chlorine, and potassium ions in the synapses, and long-term learning is the result of a feedback loop from them to the somatic genes in the axons that build new synapses between neurons. There is presumably no intentionality, content, aboutness involved in classical conditioning of the sea slug. Kandel went on to show that explicit, declarative memory in the mammalian hippocampus—first in rats and then humans—is exactly the same in its molecular biology and somatic gene switching. The only difference between what goes on in the sea slug, the rat, and the human is many more neurons are involved in the rat and even more orders of magnitude more neurons in the human. (Cf. Kandel et al. 2000.) If, as Leibniz said, *natura non facit saltus*, nature makes no jumps, if there is no original intentionality in the sea slug, there isn't any in the rat or the human. It's just an overlay we "decide" to confer on ourselves. (And the rest of nature as well. That's where religion comes from.)

A third pillar of scientism's confidence in eliminativism comes from the research program of teleosemantics. Darwin showed that all appearance of purpose in nature is really the product of blind variation and natural selection (or less misleadingly, environmental filtration). Since human behavior, including linguistic behavior, is highly purposive in appearance, the thought that controls it must be highly purposive in appearance. Accordingly, it has to be the result of, and also to be operated by, Darwinian processes of development and Darwinian mechanisms in the developed brain. This is simply because the only source of the appearance of purpose in nature is the process Darwin discovered. Teleosemantics is the research program in philosophy of mind that seeks to convert this insight into an analysis of the original intentional content of neural circuits and the derived intentional content of speech and writing. It has to work if any theory can, since Darwinism about purpose is the only game in town. Now, teleosematics gives us a lot. It fully and completely explains why we are so prone to take the intentional stance (no wonder—Dan Dennett [1969] was among the first developers of what came to be known as teleosemantics): teleosemantics accords content to neural circuitry by reverse engineering it from the marvelous environmental appropriateness of the behavior the neural circuitry produces (cf. Millikan 1984; Dretkse 1995).

The trouble is that we know in advance why teleosemantics will not give us as much or even the kind of content we need. The details behind this conclusion are to be found by considering the problems of which distal object (object of perception in the environment) is the specific topic of a neural circuit, what the difference is between mistaken content and disjunctive content, and how natural selection can discriminate nomically coextensive properties. It won't do to suggest that neural states needn't contain unique propositions and can do the work by being "ambiguous" or "ambivalent" about indefinitely many different propositions about distinct coextensive properties, since all the propositions in these sets can be united in a disjunctive, unique proposition. (For further discussion, see Fodor 1990; Fodor and Piattelli-Palmarini 2010.)

Now, when added to the Watson argument and Kandel's discovery, the right inference to draw is not that teleosemantics is a dead end as a theory of how the neural circuitry can have original intentional content. Rather, it reveals that there is no content there, just the neural causes of finely tuned behavior. Scientism urges that we draw the conclusion that original intentional content is an illusion, a myth, a convenient fiction, not a fact about us or the world, since it cannot be fixed by the physical facts.

Of course, it's owing to the lack of original intentionality and the indeterminacy of content that the interpretive social sciences and the humanities fail to provide improvable explanations of the sort that scientism will count as knowledge.

One alternative not worth canvassing for scientism is the notion that original intentionality is in public language, speech and writing, from which the intentionality of thought is derived. What, after all, could tokens of speech or inscription be but just acoustical disturbances in the air or groves or stains on clay, animal hides, and eventually paper? These physical things can no more have *aboutness* intrinsically than Watson's integrated circuits. Their *aboutness*, their intentionality, arises from the roles they play in the thought-controlled interaction of human beings. Doubtless, as a matter of genealogy, the emergence of language—nouns, verbs, adjectives, and adverbs—came back into the mind to improve thinking. (We are, after all, in Dennett's [1995] terms, Gregorian creatures, not just Popperian ones.) But the mental versions of word- and sentence-tokens are also just tokens flitting across consciousness with no intrinsic intentionality either. It is the web of connections between these tokens in conscious thought and with the spoken/written public tokens of language that builds up the illusion of intentionality, along with the apparatus that enables us to articulate that illusion. There is no language without prior cognitive agents, and all language needs to get started is cognitive agents' use of gestures and noises to facilitate cooperation. Intentionality can't get its start anywhere but in the mind, and it stops the moment minds cease to exist.

So what is the problem for scientism? Well, no original intentionality, no derived intentionality. No derived intentionality, no meaning, in particular no linguistic meaning, no semantic values, no truth or falsity. It turns out that scientism's problem of mathematical truth is the tip of the iceberg, or small beer, just a fly in the ointment, a mere technicality compared to the more serious problem that scientism cannot even express itself as a set of claims *about* anything—us, reality, ethics, and so on. That's because there is no such thing as *about*. Scientism refutes itself out of its own mouth in a sort of pragmatic contradiction, something like "I believe that there are no beliefs."

What can scientism do about this problem? Well, to begin with, it should not be treated as a counsel of despair but rather as a problem in the research program of scientism, indeed of neuroscience. And there are some promising lines of inquiry here.

Go back to neuroscience and what it is teaching us about the nature of cognition. Neuroscientists use the word *representation* to identify the neural circuits' encoding of inputs from the peripheral nervous system in, for example, the lateral geniculate nucleus or the visual cortex. But of course they use these terms, not only without commitment to intentionality, but with an explicit commitment to describing the representations in terms of structures of neural axonal discharges that are *physically isomorphic* to the inputs that cause them. Suppose this way of understanding representation in the brain is preserved in the long-term course of research, providing an understanding of how the brain processes and stores information. Then there will be considerable vindication for the mind/brain as a neural network whose physical structure is identical to the aspects of its environment that it tracks. And its representations of these features consist in this physical isomorphism. Put crudely, the representation that motion is governed by Newton's laws consists in some data structure in the brain that is isomorphic to the facts about the world that (approximately) realize these relationships. It is important to emphasize that this isomorphism between mind/brain and world is not a matter of some relationship between reality and a map of reality stored in the brain. Maps require interpretation to be about what they map, and both eliminativism and neuroscience share a commitment to explaining the appearance of *aboutness* by purely physical relationships between informational states in the brain and what they "represent." If we are going to explain intentionality, we can't sneak it in, not even a little bit. If we are going to explain away intentionality, the same thing goes. For this reason, the mind/brain-to-world relationship must be a matter of physical isomorphism—sameness of form, outline, structure that doesn't require *interpretation*.

If this approach pans out, the next task is employing it to provide some alternative for semantic evaluability, to substitute for saying that sentences or statements are true or false. Then we can hope to salvage enough semantic evaluability to ward off the suspicion that scientism cannot even be thought.

Exploiting structural isomorphism to build something like a concept of truth exploits the correspondence theory, of course: P is true if and only if P corresponds to facts about the world. Everyone knows the difficulties that attend this formulation. Many of them are circumvented by the formula that "representations"—charges at nodes in the neural circuitry—can to greater or lesser degrees be physically similar to distributions of matter and fields of energy in the environment. Which ones are more isomorphic than others? The ones that work together with the motivational components of the brain to meet needs and avoid harms, increase efficiency and reliability in these tasks, and otherwise shape behavior in ways that make it more environmentally appropriate for creatures subject to natural selection.

One has to be careful in formulating the isomorphism substitute for truth to avoid smuggling intentionality back in. For example, you can't employ wants along with (biologically, neurologically identifiable) needs to specify which neural structures are more physically similar to distributions in the world. You cannot succumb to the temptation to treat the success of some distributions of charges in the brain as *constituting* their greater accuracy than others. That would be pragmatism. Success is an indicator, not constitutive of accuracy. The substitute for truth is not success but physical isomorphism. From this notion we can also build substitutes for the full truth about reality, for full truth about aspects, components, regions of reality, as well as approximate truth, substitutes for more or less approximate truth, and substitutes for degrees of falsity.

More successful brain structures, ones more physically similar to distributions in the world, will get packaged together and sometimes result in the imprecise, illusion-ridden, but highly convenient medium of speech—silent in the mind, and noisy on the tongue.

One attraction of this approach is its friendliness to mathematics and its ability to accommodate so much of what we want by way of a fictionalist approach to mathematical statements. Physical isomorphism is often (perhaps always) reflected in sameness of mathematical structure. Insofar as we acquire physical understanding of the world, that understanding is expressed in equations and formulae. It is tempting to suggest that these mathematical structures are "represented" in the brains of the scientists who understand and employ them as accessible isomorphisms to aspects of reality, aspects we can independently "identify" by observing the behavior (including the linguistic behavior) of the scientists and their coworkers. Meanwhile, when we express these structures in sentences that appear to express our commitment to their truth, we do not utter truths (since there are no truths of any kind, really), we offer utterances and inscriptions that are part of a practice, like chess, for example, but one in which "good" moves have much greater payoffs.

It will be relatively natural to see how all the mathematics we "want" can be developed simply by the recursive rearrangement of mathematical structures in one set of neural circuits by other neural circuits. And the systematicity of these processes may well make for the systematicity of language, not just the language in which we express mathematics but the language we use for so many other transactions in life.

Of course, any two physical systems may share many different mathematical structures in common. Moreover, a given mathematical structure represented in the brain may be isomorphic with many different structures in reality. But we cannot discriminate the one it is intended to represent, or that it is supposed to be true "of." These locutions are heavy with just the intentionality that scientism denies itself. Here is a problem of underdetermination or holism that scientism shares with intentionality-dependent theories of mind. Here again, we can only invoke pragmatic criteria for discriminating successful structural representations—our substitute for true ones, from unsuccessful ones—the ones we prescientistically would call the false ones.

Scientism needs to find an alternative to communicating its claims by semantically valuable means if neuroscience does so. As is often the case in science, disciplines advance even when their serious foundational problems go unnoticed. So it is with the sciences of the brain, helping themselves to notions like "representation" that they really don't have any right to, while making significant progress in coming to grips with other more tractable problems.

Meanwhile, it is worth noticing that many of the conclusions scientism advances that are most threatening to the "manifest image," the commonsense understanding of ourselves, ultimately do not hinge on the denial of intentionality's foundations: the denials of free will, of the existence of an enduring self, of any real teleology anywhere in nature or human history, or even in individual human action. The domains of human affairs in which eliminativism makes a difference are narrative history, the literary humanities, hermeneutics, phenomenology, psychoanalysis, and pop psychology. The claims of all these disciplines will turn out not even to be false, if there is no such thing as meaning at all.

How can these theses of scientism be understood without being expressed in ways that require intentionality, the meaningfulness of thought and speech in terms of content and *aboutness*? Well, roughly, scientism turns out to be something like this: once the human brain begins to register neural circuitry isomorphic with all or sufficiently large regions of reality, it will be able to determine from these neural structures alone (and without further search for new structures isomorphic with reality) that the reality to which they are isomorphic contains no structures even resembling those required by the manifest image, common sense, religion, or other wishful thinking. Of course, as David Hume advised, we'll still be able to speak with the vulgar while we think with the learned.

......

Literature Cited

Balaguer, M. 2011. "Fictionalism in the Philosophy of Mathematics." *The Stanford Enyclo-pedia of Philosophy* (Summer 2015 ed., with minor correction), edited by Edward N. Zalta, http://plato.stanford.edu/entries/fictionalism-mathematics/.

Chalmers, D. 2009. "Ontological Antirealism." In *Metametaphysics: New Essays on the Foundations of Ontology*, edited by D. Chalmers, D. Manley, and R. Wasserman, 77–129. New York: Oxford University Press.

Dennett, D. 1969. *Content and Consciousness*. London: Routledge.

———. 1995. *Darwin's Dangerous Idea*. New York: Simon and Schuster.

Dewey, J. 1938. *Logic: The Theory of Inquiry*. New York: Holt, Rinehart and Winston.

Dretske, F. 1995. *Naturalizing the Mind*. Cambridge, MA: MIT Press.

Field, H. 1980. *Science without Numbers*. Princeton, NJ: Princeton University Press.

Fodor, J. 1990. *A Theory of Content and Other Essays*. Cambridge, MA: MIT Press.

Fodor, J., and M. Piattelli-Palmarini. 2010. *What Darwin Got Wrong*. New York: Farrar, Straus and Giroux.

Gould, S. J., and R. C. Lewontin. 1979. "The spandrels of San Marco and the Panglossian paradigm: a critique of the adaptationist programme." *Proceedings of the Royal Society of London. Series B. Biological Sciences* 205 (1161): 581–98.

Hirstein, W. 2006. *Brain Fiction: Self-Deception and the Riddle of Confabulation*. Cambridge, MA: MIT Press.

Horgan, T., and J. Tienson. 2002. "The Intentionality of Phenomenology and the Phenomenology of Intentionality." in *The Philosophy of Mind*, edited by D. Chalmers, 520–32. New York: Oxford University Press.

Kandel, E., T. Jessel, and G. Schwartz. 2000. *Principles of Neuroscience*. New York: McGraw-Hill.

Kitcher, P. 1984. *The Nature of Mathematical Knowledge*. Oxford: Oxford University Press.

Kornblith, H. 2015. *A Naturalistic Epistemology: Selected Papers*. New York: Oxford University Press.

Lewis, D. 1993. "Mathematics Is Megethology." *Philosophia Mathematica* 3: 3–23.

Loftus, E. 2008. *Eyewitness Testimony: Civil and Criminal*. 4th ed. Charlottesville, VA: Lexis Law.

Millikan, R. 1984. *Language, Thought and Other Biological Categories*. Cambridge, MA: Bradford Books.

Purves, D., G. Augustine, D. Fitzpatrick, W. Hall, A.-S. LaManti, and L. E. White. 2012. *Neuroscience*. Sunderland, MA: Sinauer.

Quine, W. 1969. *Ontological Relativity and Other Essays*. New York: Columbia University Press.

Rorty, R. 1981. *Philosophy and the Mirror of Nature*. Princeton, NJ: Princeton University Press.

Rosenberg, A. 2006. *Darwinian Reductionism; or, How to Stop Worrying and Love Molecular Biology*. Chicago: University of Chicago Press.

———. 2011. *The Atheist's Guide to Reality: Enjoying Life without Illusions*. New York: W. W. Norton.

Rosenthal, R., and L. Jacobson. 1968. *Pygmalion in the Classroom*. New York: Holt, Rinehart and Winston.

Salmon, W. 1989. *Four Decades of Scientific Explanation*. Pittsburgh: University of Pittsburgh Press.

Searle, J. 1980. "Minds, Brains and Programs." *Brain and Behavioral Science* 3 (3): 417–57.

Shapiro, S. 2000. *Thinking about Mathematics: The Philosophy of Mathematics*. New York: Oxford University Press.

Susskind, L. 2014. *The Theoretical Minimum*. New York: Basic Books.

van Fraassen, B. 2002. *The Empirical Stance*. New Haven CT: Yale University Press.

Weinberg, S. 2015. *To Explain the World*. New York: Harper Collins.

Weiskranz, L. 2009. *Blindsight*. New York: Oxford University Press.

Yablo, S. 2009. "Must Existence-Questions Have Answers?" In *Metametaphysics: New Essays on the Foundations of Ontology*, edited by D. Chalmers, D. Manley, and R. Wasserman, 507–26. Oxford: Oxford University Press.

12

Economics and Allegations of Scientism

DON ROSS

Introduction: Varieties of Scientism

There is no fact of the matter about what scientism is, only facts about what different sets of people take scientism to be; and as the chapters in the present volume make clear, there is much disagreement across these sets. We can imagine an imperialist scientism that aims to promote the value of science at the expense of other human activities that have quite distinct purposes, such as art. Such scientism would be foolish and pointless; although governments and other institutions must frequently make trade-offs between funding science and funding other activities, they should never do so on the basis of decontextualized value ranking on a grand and general scale of comparison. The general public is probably most interested in the form of scientism that resists the claims of religious authorities to have access through faith and scripture to esoteric objective truth. With this sort of scientism I am entirely in sympathy, but it is a live issue mainly in the competition of cultural ideologies rather than in scholarly philosophy. A more refined version of this scientism is invoked by Ladyman and Ross (2007) (L&R) against so-called analytic, or a priori, metaphysics. Both the popular and the refined versions apply the proposition that science enjoys a monopoly among institutional practices that are well designed or well adapted to discover generalizable objective empirical truths.[1] L&R's explicit critical target is metaphysical speculation that is not based on, and is indeed often willfully disconnected from, science.

The claim that objective truths, at least outside mathematics, are better sought through empirical science than through analytic metaphysics or intuitive reflection has little clear force in the absence of some indication of how empirical science is distinguished from these other practices. This is not straightforward, because scientists sometimes analyze concepts,[2] and exercise intuitions in judging which theoretical structures are best applied to which phenomena. In the philosophy of science, study of this demarcation problem has usually focused on putative *methodological* characteristics that allegedly distinguish science from nonscience. L&R (2007) reject this approach on grounds that there is no such thing as generic scientific method, though we recognize that there are some epistemological principles that all scientists are expected to follow, such as controlling for sample selection biases when making inferences about populations from samples. These are not special scientific principles, however, because no one advocates their suspension in *any* domain. When L&R (2007) defend the preeminence of science in seeking objective generalizable truth, what they mean by *science* is an *institutional* structure that perpetuates itself by rewarding certain behaviors on the part of its participants and filtering out other behaviors. Scientific journals, doctoral examination boards, tenure committees, prize judges, and other adjudicators do not, unlike analytic metaphysics journals[3] (etc.), reward preoccupation with the meanings of folk concepts, or direct reliance on intuitions that are not buttressed by systematic observation.

Pigliucci (this volume) comes to a similar conclusion in reflecting on demarcation. He also notes, correctly, that this implies a fuzzy frontier, because institutional similarities and dissimilarities between practices are always matters of degree. This doesn't bother him because, he says, we can distinguish clear pseudosciences such as astrology that are far on one side of the boundary from clearly scientific practices, such as astronomy, that are equally firmly within the other side. But the demarcation problems that have actually mattered, in the sense of arousing genuine controversy within the academic establishment, are precisely those that occur along the fuzzy borderlands. These have typically been cases where people apply methods characteristic of well-established sciences to new subject matters, where they encounter and excite resistance from resident authorities who have been studying the subject in question in a different way. Methodological demarcation efforts have often treated physics, chemistry, and biology as paradigmatic sciences, from which the essential characteristics of science in general can and should be inferred inductively. This in turn has led naturally to contestation around the scientific bona fides of nonparadigmatically scientific disciplines, particularly those that study social phenomena. Thus, for example we find Winch (1958) arguing that "social science" is an oxymoron, and, by exact contrast, Papineau (1978) identifying and promoting methodological and ontological principles by which social inquirers can and should be "more scientific."

When this debate is pursued on the *general* plane, one side of it—Winch's—is in the position of Isaiah. The social sciences are a formidable cluster of institutions that are not about to be argued out of existence by philosophers. This is part of the *reason* why L&R (2007), as naturalistic philosophers, favor institutional over methodological, epistemological, or ontological demarcation of science; according to them, philosophers have no authoritative ground on which to stand and pronounce doom on the social sciences without appearing ridiculous, and if a proposed demarcation criterion had the consequence that the social sciences in general are not real sciences, this should be a basis for rejecting the criterion in question. Social science is carried out on a large scale, by scientific institutions, and so a task for the suitably humble philosopher of science is to seek interesting generalizations about how it is done and why, and a task for the (naturalistic) metaphysician is to explore the implications of its practice and results for the general structure of the world (i.e., to consider the nature of generalizations about historically restricted patterns, the focus of L&R's [2007] fourth chapter).

This cannot be the end of the matter, however. The argument above rests on the idea that naturalists are obliged to treat any inquiry that is taken institutionally seriously by practitioners of empirical inquiry as prima facie deserving of philosophical attention. But then we cannot dismiss as unimportant the fact that *within* each of the disciplines of social inquiry we find serious debates that are appropriately construed in terms of critiques of scientism. When a historian defends classical narrative inquiry as (typically) more enlightening than cliometrics, or a cultural anthropologist promotes the method of *verstehen* as superior to statistical analysis of observed data, it is not unreasonable to regard these as situated quarrels over scientism, since they turn on the extent to which the governing goal of the discipline in question is *generalizable* truth.

I stress "situated"; a key characteristic of these sorts of engagements within disciplines is that they are actually about the historical record, or actually about the sources and significance of cross-cultural and intercultural variation; they are not just philosophical arguments carried on by reference to examples. Thus, although they are the sites of genuinely consequential arguments over scientism, they are not, in the first place, motivated by questions about whether scientism in general is a good or bad attitude. This is also why I referred to debates within *each of* the social inquiry disciplines rather than to *a* debate across *all* of them. There is nothing inconsistent about doubting the achievability or value of broad generalizations in (for example) history while embracing their pursuit by, say, sociologists.

It would be strange to assert that we find *distinct* conflicts around scientism in each social inquiry discipline, and then proceed to write a single short essay about all of them. So I will not do that. I will instead focus on a single case study, of rhetoric about scientism in economics, that is particularly interesting in part because economics is the discipline in which scientism is most often alleged to lead to problematic policy and social consequences (Dupré 2001).

A Century of Scientistic Rhetoric in Economics

It is not uncommon for activists and governments that try to promote "progressive" agendas to be chided—at least by economic journalists if not so much by actual economists—for trying to evade or overturn "the laws of economics." Such rhetoric infuriates critics like Dupré (2001), who explicitly identifies it as a form of oppressive scientism on grounds that it tends naturally to shore up the status quo. It is scientism, according to Dupré, because there in fact are no laws of economics. On the other side of the ideological ledger, golden-age socialists of the Soviet (as opposed to the Frankfurt) style used to characterize their economic program as "scientific." The refiners of socialist economics, Oskar Lange and Abba Lerner, took themselves to be discovering objective general facts when they identified national production frontiers, and insisted that central planning that deliberately targeted the frontiers in question could reach them and avoid the volatility associated with less regulated markets. *Scientific* in this instance evidently meant: able to furnish a basis for engineering. Hubris about what can be engineered is yet another thing to which scientism can sometimes refer, and very often is intended to refer by those who use the concept.

It is not surprising that what socialists took to be *distinctively* scientific in their program was its applied, rather than its theoretical, aspect; otherwise, the intended contrast class would have been entirely unclear. It is hardly an exaggeration to say that between the rise of economics as an institution in the 1870s until the past couple of decades, all leading economic theorists characterized themselves as producing science. Mirowski (1989) has argued persuasively that various important features of early marginalist economics—the body of thought that launched economics as a recognized autonomous academic discipline—were imported directly from the modeling framework of classical mechanics on the basis of no reflections about or observations of economic behavior, but simply because classical mechanics was the exemplar of scientific thought. Specifically, Mirowski argues that this explains why economics became structured around applications of calculus, why economic agents were modeled as maximizers of (utility or production) functions, and why the characteristic result of an exercise in economic analysis was identification of an equilibrium. More tendentiously, Mirowski contends that the classical mechanical framework failed to adequately fit the target domain because there was no empirically motivated principle of market structures that was analogous to the classical mechanical principle of energy conservation.

Mirowski exaggerates the extent of bad faith involved in this borrowing because of his own preoccupation with counterscientistic rhetoric. There was much more to early marginalist economics than hijacked analytical methodology, and also more heterogeneity of views among the founding generation than Mirowski credits. For present purposes, however, we can concede that like their contemporaries in psychology and the Weberian wing of sociology, the late nineteenth-century economists were certainly self-conscious and insistent about claiming the status of scientists, and that they took literal resemblance to classical mechanics as the hallmark of the scientific. Among the most important problematic legacies of this attitude that has haunted mainstream economics ever since—though it has seldom been formally encapsulated in core theory and has often been disavowed when critical pressure has been brought to bear on it—is atomism. This is reflected in the empirically unmotivated tendency of economists to represent market structures as compositions of self-subsistent individual agents with intrinsic, asocial preferences and capabilities. This criticism must be balanced by noting that the single most important principle of the modern discipline, downward-sloping aggregate demand for bundles of substitute consumption streams,[4] was later shown theoretically and empirically not to require atomistic decomposition, and to be derivable at the aggregate scale simply from the assumptions of constrained household budgets plus increasing heterogeneity of consumption with increasing incomes (Hildenbrand 1994). This presages a main point of the current chapter: when we discover that some rhetoric has been adopted mainly because of its association with highly valorized parts of science, we have not thereby shown that this significantly undermines the substantive analysis that the rhetoric decorates. Arguments over charges of scientism, even when made rigorously, typically don't establish much of importance.

If economists during the Victorian period and the first half of the twentieth century were no *more* likely than other practitioners of young disciplines to establish status by pinning "scientist" badges on their lapels, this changed immediately after World War II. Just when (some) psychologists, sociologists, and anthropologists began to take hermeneutic turns, or at least admit the value of "qualitative" research approaches that took them closer to the narrative humanities, economists, emboldened by the valorization of— and genuine successes of—quantitative operations research during the war lurched sharply in the direction of institutionally enforcing a very specific, narrow philosophy of science. The key text, as acknowledged by all historians of economics, was Paul Samuelson's *Foundations of Economic Analysis* (1947).

Though I have written elsewhere about Samuelson's philosophy of science, and about his influence on the methodological norms of his discipline (Ross 2005; 2014), I approach this now with special caution. The historian of economics Roger Backhouse is soon to publish what will be the authoritative intellectual biography of Samuelson, and in the nature of such contributions it is sure to overthrow some comfortable generalizations. So it is risky to write things like "It is clear that Samuelson intended *x*" unless Samuelson very explicitly said "*x*." I am not aware of Samuelson having literally announced that he had rendered economics truly scientific, by his lights regarding what "truly scientific" meant. But, going out an a limb with Backhouse coming up the tree, I think it would be difficult to deny the accuracy of that gloss of both Samuelson's original 1947 introduction to the *Foundations*, and his 1983 introduction to the enlarged edition. His artfully oracular style allows him not to seem to boast when, in both introductions, the main figure with whom he aligns himself historically is Newton. On page 1 of the 1983 text, he places himself among those "who consider science the most exciting game in the universe."

Samuelson thought he had made economics scientific by reconstructing what he took to be the sound economics of his day, including quite a bit he had built himself, on the basis of "operationally meaningful theorems." By *meaningful* he meant "falsifiable," and what rendered the theorems falsifiable, according to him, was their precision. This was achieved through stating them mathematically rather than "verbally," to use the standard, if slightly weird, word that economists still use to mean "not mathematically."

Sociologically, Samuelson's methodological intervention was utterly victorious. Before the 1950s, most articles in leading economics journals were primarily "verbal." Since the mid-1960s onward, almost none have been (though in 1987 the American Economic Association created the *Journal of Economic Perspectives* as a sort of red-light district where "verbal" expression is allowed, provided it is backed up by relevant mathematics available elsewhere).

In what might be claimed to be its most basic meaning, *scientism* is used to refer to attempts to impose what are taken to be norms that govern scientific practice on domains where the norms in question did not formerly apply. On that usage, Samuelson can be held up as a veritable icon of scientism. But then it is open to the critic to ask whether the norms being extended really are scientific ones, or rather philosophical strictures that reflect a parochial *image* of science.

Samuelson's emphasis on falsifiability has fed a strongly Popperian theme in the (limited) methodological literature that was published by economists during the period between 1960 and (roughly) 1985; later this was supplemented by an emphasis on the revision of Popper by Lakatos, who added the comforting idea that good science could include a "hard core" of assumptions that are not directly subject to empirical test. Blaug (1980) influentially stated the Popperian rhetorical orthodoxy, while also sternly taking his fellow economists to task for paying little more than lip service to it. Almost certainly, economists' widespread lack of interest in the truth of core assumptions during the period under discussion was more directly influenced by Milton Friedman's (1953) famous argument to the effect that the truth of assumptions in economics is irrelevant if they are used to generate accurate predictions.[5] But Friedman relies on no coherent philosophy of science, as methodologists generally recognized; for them, Lakatos was available as a fallback buttress to practice.

Economists' reputation for not caring about actual empirical data, at least until lately, is overblown. What is true is that almost all empirical work in economics tests statistical estimations of models constructed and interpreted under the guidance of theories; economists often aim to test models informed by theory, and almost never directly test background axioms. Recognition of this explains the persisting popularity of Friedmanian rhetoric among economists and Lakatosian philosophy among methodologists, though methodological apologists would earn more respect from philosophers of science if more of them instead followed the lead of Morgan (2012) in attending to philosophical work on models and their role in connecting theoretical structures with reality. They would likely then more widely recognize that their practices with respect to theory testing are not markedly different from those of investigators in more paradigmatically scientific disciplines who study highly stochastic, noisy systems. The real problem for would-be Popperian economists for a few decades after Samuelson was that until the spread of game theory in the 1980s allowed economists to model imperfect markets, they lacked the tools to specify the partial equilibria of real, semi-isolated markets (Sutton 2000; 2001); and until the 1990s they lacked the computational power or the statistical technology to often perform the identifications on which model discriminations that matter really turn. But the second point is equally true of macro-scale biology. The situation is utterly changed now, for biologists and economists alike.

Thus, we might summarize the high-level history of scientistic rhetoric in economics until recently as follows. The founders of the modern discipline, like their peers in the behavioral and life sciences generally, were concerned to establish the legitimacy of both their institutions and their claims to knowledge as being properly scientific. Anyone setting out to do this must do so on the basis of some philosophy of science or other. During the first period in its history when economics settled into a view that clearly counted as orthodoxy, immediately after World War II, the philosophy of science that was pulled off the shelf was falsificationism. In retrospect, this was not a happy choice. Though implementing a falsificationist program does require reasonably precise claims, as Samuelson and his followers emphasized, it also requires tractable empirical measurement protocols; and in economics these remained very immature until the late twentieth century. Worse, falsificationists tended to be naïve about the relationship between theories and their empirical substructures, which is particularly problematic in an enterprise that is as reliant as economics on building models of partially isolated aspects of reality (Rodrik 2015). Economists thus made themselves vulnerable to charges of illegitimate scientism because their rhetoric laid claim to scientific status by extolling methodological virtues they did not appear to practice. But then the problem turned out to rest on technological limitations rather than bad faith. As the computational power available to them grew, economists lavished ever more relative attention on building more sophisticated econometrics that could allow them to estimate models of data; and now the overwhelming majority of papers published in the top economics journals consist of such estimation exercises. The claim that economics is scientistic, in the pejorative sense, now looks highly undermotivated.

Because there is no grand consensus in the philosophy of science, a critic always has the option of conceding that economists might now rise to *their own* standard for scientific practice, but insisting that this standard isn't high enough. For example, it might be alleged that econometric techniques fail to adequately address the identification of causal structure (Kincaid 2009), and that that is an essential ambition of "genuine" science. However, this charge can be countered by citing much evidence of concern to establish causal relationships in economics, through design of randomized experiments (Banerjee and Duflo 2011), use of instrumental variables in econometric estimation (Angrist and Pischke 2009), and openness to consideration of hypothesized latent forces in structural models (Andersen et al. 2010). In any case, even if it can still be said that many economists are underattentive to the complications of causal inference, it doesn't seem that any force is added to the criticism by complaining about scientism.

If this were the end of the story, then it wouldn't appear that the case study of economics adds much enlightenment to general discussions of scientism. The case study, I would argue, is representative of the fact that allegations of scientism typically require assumption of some normative philosophy of science that is unduly restrictive. Science is nothing more or less than a set of institutions for trying to discover objective truth, and there can be nothing inappropriate about applying such institutional structures to any domain where objective truth might be found.

However, we have *not* yet surveyed the whole story where scientism and economics are concerned. Recently, some leading economists have promoted rhetoric suggesting that decades of allegations of scientism have drawn blood. These economists seek to *defend* economics by *distancing* it from "pretensions" to scientific status. The moral I will draw from review of this situation is that fussing over scientism is not merely harmless but pointless distraction, because it can encourage scientists to adopt wrong-headed philosophies of science in self-defense.

Evading Allegations of Scientism

The global financial crisis that began in late 2008 revivified doubts about economists' claims to knowledge, and hence about the status of economics as a science. Economists, it was often said, failed to see the collapse of the US housing market coming, and failed to see that that market was a bomb that could blow up the global banking system. I have argued elsewhere (Ross 2010) that this perception is inaccurate as applied to most economists. I will not revisit that debate here, as my present subject is scientism, not economic forecasting. But when I want to show a skeptic my favorite instance of pre-crisis wisdom about financial downturns, I refer them to Edward Leamer's (2008) gripping study of the history of the US business cycle since World War II, written for nonspecialists (specifically, his MBA students). Through careful sifting of data, guided by no exotic or overgeneralized model but with a clear alertness to the multiple interacting network of causal relationships in the national-scale market for goods, services, real estate, and financial assets, Leamer identifies the indicators of looming trouble. These indicators were flashing red well before the fall of Lehman Brothers, and most economists of my acquaintance thought so. This was picked up by the mainstream economic media, and given repeated front-page coverage in, for example, the *Economist*, illustrated by cartoons of houses falling on Wall Street.

The example of Leamer's 2008 book is in my view a clearer guide to economic methodology than any explicitly methodological tract of which I am aware. But in the wake of the hyperbolic expressions of doubt about the health of the discipline that followed the crisis, a number of economists produced such explicit reflections. Leamer (2012) was among them. Unlike some others—for example Coyle (2007) and Rodrik (2015)—Leamer does not try to defend the relevance and expertise of economists by surveying the extent of sound foundations in the discipline's several branches. He instead focuses on a single venerable relationship, represented by a family of simple models known as Heckscher-Olin, between the factor endowments of countries (or regions) and their export and import profiles. The basic model, he explains, is much too simple to be literally true, and economists have produced variations and refinements on it that are adapted to different kinds of circumstances. Policy-relevant insight is achieved into such questions as the effect of more open trade on wage levels by regarding the model not as the one correct, overarching account of international or interregional trade, but as a source of discipline on thought for people who work with it against a backdrop of close familiarity with data on large-scale production, consumption, and wage setting, and the nature of decision making in business. The H-O model, Leamer aims to show, is an elegant *tool* that is essential for best policy advising in the hands of a practiced economic *craftsperson*.

The relevance of this to our present reflections on scientism is that the contrast class for the economist's "craft" is taken to be science. The same distinction plays a central role in a more recent book-length defense of economics, that of Rodrik (2015). It will be recalled that a standard move in diagnoses of scientism is to begin with a characterization of a paradigmatic science or body of scientific practice—often physics—and then find that an alleged extension of the paradigmatic practice (or epistemology, or methodology) fails to be appropriate to the new domain of application. In what follows, I will observe tradition in taking physics to be the uncontroversial exemplar of science applied properly to appropriate phenomena.

In striking contrast to the Samuelsonian tradition reviewed in the previous section of the present chapter, Leamer's rhetorical strategy is in effect to defend economics against charges of scientism by arguing that the best economics *does not aim to be science*. According to Leamer, once we properly understand economics as a craft, we can begin to distinguish better, more careful and insightful, economics from misleading and overambitious economics in the same way that we can distinguish (say) careful and intelligent from reckless and inattentive carpentry or baseball pitching.

To support his science/craft dichotomy, however, Leamer must suppose that economics is markedly different from physics in its aims. He argues that the economist's primary goal is not to produce models that *accurately* describe the world; rather, he insists, it is first and foremost a practical activity intended to improve policy. According to Leamer, a theoretical model such as H-O is "a simplified and distorted model of the real world." By contrast, it is generally held that what distinguishes engineering from physics is that physicists aim to discover the true character of reality regardless of whether it is useful. Leamer further maintains that a necessary condition for good economic modeling is that it reveals insightful stories. By contrast, there is a strong tradition in commentary on physics that explicitly denies that physicists should tell stories.

On the alleged contrast between economic and physical models, we turn to the best work by philosophers of science who have studied physical models. Cartwright (1980; 1983) and Morgan and Morrison (1999) provide example-rich accounts to argue that models in general *isolate* small sets of relationships from the complex causal weave of the world and try to make generalizations about channels of influence propagation—that is, general causal connections—in these simple pretend worlds. In a typical physical experiment, the scientists *literally* isolate the special system by shielding the lab. When we worry about whether a physicist's model is true, we should mean "true of the isolated system," not "true of the uncontrolled world." Although he sometimes speaks of economic models as "fictions," at other points in his account of the H-O model Leamer likewise focuses on what is true *in the model*. For example: in an H-O model it's *false* that, contrary to conjectures based on some rival models, volumes and values of imports drive changes in wage levels. We can use the model, plus econometrics, to try to determine whether or not this relationship might sometimes prevail in an actual economy.

What motivates Leamer's concern to distinguish science from craftwork, and thus to exaggerate the differences between physical and economic models? As noted earlier, people often become disenchanted about economics because it fails to predict the timing of large-scale economic events; and this disappointment has been particularly manifest since the onset of the 2008 financial crisis, from which global recovery has been slow and incomplete. People also become skeptical about economics because they observe a proliferation of models for characterizing relationships among the same general variables. Most are unaware that, as Cartwright (1980) illustrates with the example of quantum damping (the loss of a proton when a charged particle passes through a magnetic field), the same point applies to physics; physicists deploy six mathematically divergent models of damping, depending on experimental context, although all the models agree on the basic causal relationship. But many people think that "real" sciences are essentially deterministic prediction machines, and don't generally grasp the differences between general theories and specific models. I conjecture that Leamer, wishing to defend his accurate conviction that he and his colleagues have genuine expertise in identifying economic causes of events, but aware that they don't apply this expertise by simply applying theoretical algorithms, implicitly buys into the popular misconception of paradigmatic science when he characterizes their advantage as skill rather than knowledge.

A particularly illuminating part of Leamer's book is his criticism of economists who fail to distinguish between "mere" theory—that is, extensions of axiom sets without regard to whether they're ever likely to model any real economic circumstances—and use of theory informed by rich historical, empirical experience to build structural models that generate causal stories about real economic markets or industries or institutions. Leamer is particularly insistent—as he is in his previous book on the US business cycle—that these vehicles of understanding are *stories*.

In light of this, one philosopher of science whose work might be thought to lend support to Leamer's distancing of economics from science is Alex Rosenberg. Rosenberg (2011) distinguishes science from popular nonfiction precisely on the ground that science does *not* construct stories but instead models functional relationships that hold timelessly across broad ranges of specific episodes; and he takes physics to be exemplary science in this respect.

The aspect of storytelling that Rosenberg emphasizes in this argument is narrative structure. This aspect is not particularly important to the role of stories emphasized by Leamer, though, as he says, narratives help people *remember* economic lessons. Rather, Leamer's stories are chains of causation that pull some causal relationships into the foreground while suppressing others. Thus, they function just like models. Indeed, we could say that Leamer's stories are verbal, causal models that complement the formal, functional models he uses. They add value, not by identifying causal relationships that the models leave out, but by distinguishing some of the causal relationships as potential action points for policy interventions.

This feature of Leamer's account is also more familiar than he seems to suppose from accounts of best-practice modeling in physics that have been articulated by philosophers of science such as Woodward (2003). The formal principles of physics are generally symmetries that make no reference to—could not make reference to—causal influences. But a physical experiment is all about intervening in a simplified system to *make something happen.* According to Woodward, what it *is* for something to be a physical cause is to be the possible basis for a difference-making intervention; and this is currently the most popular account of causation among philosophers of physics.

A remark of Leamer's that would certainly not be endorsed by many contemporary philosophers of science—though the positivists of the previous century would have been happy with it—is "Correlations are in the data, causation is in the mind of the interpreter." Philosophers are likely to read this out of context as an affirmation of antirealism about causation. But it is quite inconsistent with Leamer's general philosophical modesty that he might intend to be advancing a metaphysical assertion by this remark. His point, rather, is that the causal structure of a whole economy is so complicated that if we tried to present it we wouldn't be able to understand it, and certainly wouldn't be able to design better policy on the basis of it. The act of "interpretation" consists in foregrounding the parts of the causal structure that matter, and that matter recurrently.

An obvious question to ask at this point is, Why shouldn't we say the same thing about the correlations we detect? Why are they, by contrast, said to be "in the data"? I will offer a suggestion derived from reflection on physics. Physicists worry, just as Leamer wishes economists would worry more, about distinguishing mathematics from their substantive subject matter. But in physics, getting a clear hold on the distinction is harder than in economics, because the processes and objects to which the math is applied are themselves constructed from theory, and the theory is of course mathematical. Is a quantum field a kind of thing or a mathematical abstraction? Such questions aren't idle philosophizing in physics. Critics of string theory, for example, complain that it's just math (Smolin 2006), and if that criticism turns out to stick, then string theorists will stop getting jobs in physics departments.

Ladyman and Ross (2013)—in a context where, unlike Leamer's book, metaphysics *is* the point—have recently argued that when physicists thought that the world was deterministic and mechanical, they distinguished mathematics from physics by reference to the idea of matter: physical systems were *material*, and they at least thought they understood what they meant by that. But for the past hundred years, what has marked the passage from mathematics to physical application is that the scientist takes a formal, mathematical theory and uses it to build a *statistical* model. Philosophically, this is a more satisfying state of affairs. *Maybe* mathematics is merely formal. But statistics aren't formal at all. Statistical relationships just *are* the data. And, given the strong possibility that quantum physics, the body of theory that describes fundamental reality, is irreducibly nondeterministic,[6] statistical relationships might characterize the most general metaphysical structure of the world. Ladyman and Ross promote the following metaphysical slogan: "The world is the totality of non-redundant statistics." If this is right, then economics is *much* more like physics than almost *anyone* has thought, at least since the Victorians, who are treated so roughly by Mirowski.

Leamer seems to think, echoing a view I have heard from other economists, that one isn't doing science if one critically analyzes lots of statistics to produce causal stories. But, I have been arguing, that is precisely what more and more scientists have been doing as the idea that the world is fundamentally mathematical gives way to the idea that it might be fundamentally statistical.[7]

Conclusion

Samuelson and his immediate successors thought that "being scientific" meant observing the following methodological recipe: axiomatically characterize a class of systems so that interpreted formal theory generates univocal predictions[8] about all members of the class. Empirical disconfirmation would thereby refute the axioms (under the relevant interpretation); in which event, one should revise them and try again. But these economists did not implement this program, and thereby exposed themselves to allegations of scientism, that is, of cloaking their work in a mantle of science that their actual practice did not support.

The charge of scientism, however, was and is idle, because it rests on a pinched philosophy of science. No one ever did empirical science in the way that Samuelson's introductory rhetoric in his *Foundations* implies. It is true that for a few years the leading lights of economics did a lot of mathematical modeling and very little empirical science. But now, with the facilitation of richer econometric estimation techniques made possible by abundant computational power, most economists are up to their shoulders in empirical science—but some have come to think that it isn't science!

This strange state of affairs arises from fear of scientism—but that in turn arises from misplaced efforts to demarcate science from nonscience on the basis of a priori methodological or epistemological strictures. Science is a set of institutional practices, which will be deployed opportunistically wherever they yield results. It might have turned out to be the case that they yield results only when universal generalizations deterministically generate predictions— but matters are *not* turning out that way, perhaps not even in fundamental physics. Having *effective* beliefs about the world—in physics, economics, biology, law, political journalism, business management, governance, and so on— largely consists in carefully and critically modeling statistical distributions of types of events and relationships, and identifying causal channels. This list includes a variety of practices that are rightly not regarded as parts of science. The differences between scientific and nonscientific approaches to problem management are institutional, not methodological.

This undermines the possible *point* of worrying about scientism. If there is no general, distinctive scientific method, then we can't soundly allege that such a method is applied to a range of domains it doesn't fit. "Bad" scientism in the sense that most users of the term intend would have to involve advocating the use of scientific institutions to govern processes not mainly aimed at growing objective knowledge. But there are almost no such advocates.

There are, to be sure, people who admire paradigmatic sciences such as physics and don't admire other activities so much, or at all. Sometimes—as in Ladyman and Ross's (2007) rejection of analytic metaphysics—admiration is withheld because the activity purports to generate effective beliefs about the world but consistently ignores relevant data and mis-models such data as it does attend to. But then the problem isn't that analytic metaphysics isn't scientific but that it isn't empirically alert to facts about the domain it aims to address. If it did seriously address these facts, then physicists, defined institutionally, would have no general grounds for dismissing it. Indeed, that is why L&R maintain that there is room for a naturalistic, as opposed to analytic, metaphysics. Whether naturalistic metaphysics is or isn't science is a question to be resolved by the future evolution of institutions and is not a question of philosophical substance.

Those critics preoccupied with scientism who don't admire anything that looks and feels "scientific" but isn't *paradigmatic* science rely, at least implicitly, on the idea that there is a distinctive, sound scientific method that is exemplified by the paradigm(s). Allegations that economists have carried out a scientistic program exemplify this attitude. Such allegations are usually normatively motivated by the feeling that economists try to falsely cloak themselves in the respect earned by physicists, through borrowing its methodological trappings. Such critics hold this to be illegitimate on grounds that the methods in question can't be made to fit the domain of application.

My response has been to argue that insofar as science has methods, they are merely those that apply to all empirical inquiry. When L&R (2007) deliberately don the mantle of scientism, this is the intended point; metaphysics had better be empirical inquiry or else be useless. But this is a special case with a special history. People who worry about scientism more typically apply the label to activities such as economics, for the reasons indicated above. But unless someone doubts that economics should even try to be empirical inquiry, this form of criticism cannot deliver its intended punches. In his sophisticated way, Leamer flirts with the idea that economics is not empirical inquiry but only useful spinning of fictions. But we can apply the rhetoric that sustains this conceit equally well to physics. The real issue that motivates Leamer, along with less friendly critics of economics, is the emptiness of some of it, which stems from confusing purely formal inquiry with empirical inquiry. That merits criticism, but it is not usefully understood as scientism.

......

Notes to Chapter Twelve

1. The inclusion of *generalizable* here is intended to exclude pursuit of facts in which interest doesn't derive from an ambition for generalization. For example, there is little point in calling the inquiry that goes on in a courtroom scientific, though it aims to discover objective empirical truths. A soldier on a reconnaissance mission is not doing science, though she seeks empirical facts. Attempts to distinguish better from worse military strategies, on the other hand, are best pursued scientifically.

2. A relevant difference here is that scientists analyze *their own* concepts, whereas analytic metaphysicians begin with concepts they find in general folk cultures, though they soon refine these beyond popular recognition.

3. Many incentive structures, not merely those in analytic metaphysics, reward features that scientific incentive structures do not. I am here singling out analytic metaphysics, as in L&R (2007), because it competes with science in a special sense: along with fundamental physics, it purports to objectively characterize general structures of the world. And unlike a number of branches of philosophy that are concerned with normative questions about how to live or how to think, metaphysics seems to resemble science in needing to aim at truth if it is to aim at anything worthwhile. Van Fraassen (2002) emphasizes this point.

4. This is a technically correct way of expressing the general idea that, as a market consumes more of a product, holding budgets and population fixed, that market's marginal demand for the product falls as other wants are attended to. Almost nothing in economic theory, micro or macro, would be well motivated if this generalization didn't hold.

5. Thanks to Ivan Boldyrev for making this point.

6. With hidden variables now demonstrably excluded from the equations of quantum mechanics, a deterministic quantum physics appears to require the many-worlds interpretation of Everett (Deutsch 2011; Wallace 2012). Ladyman and Ross (2013) argue that that is not a *reason* to prefer the many-worlds interpretation, because there is no pressing reason to assume that physics must be deterministic.

7. This idea bears directly on the argument of Rosenberg (2011 and this volume) for a scientism of a kind that every opponent of it, on any reasonable definition of it, aims to resist. Rosenberg's scientism is not as easy to deny as most of his critics think. He is right, for example, that naturalism is incompatible with belief in original intentionality (see Dennett 1987), so blocking eliminativism about intentionality altogether requires an account of how intentionality could be both derived and real. (Evidently it must be derived, at least to begin with, from something nonintentional.) This requires denying a premise Rosenberg doesn't discuss in his chapter in the present volume: ontological reducibility-in-principle of all real patterns to fundamental physics. What that denial in turn requires is recognizing that mechanics, including quantum mechanics, is not fundamental physics. The best current candidate for fundamentality in physics is quantum field theory, which isn't about matter in the sense of matter that folk materialism, and philosophical materialist reductionism, relies on. The road to that conclusion, finally, depends

on framing reality as irreducibly statistical, a suggestion that goes back to C. S. Peirce. See Ladyman and Ross 2013.

8. In referring to *interpreted* formal theory, I follow usage that goes back to the logical positivists. Uninterpreted formal theory makes *no* empirical predictions, because it is just mathematics.

Literature Cited

Andersen, S., G. Harrison, M. Lau, and E. Rutström. 2010. "Behavioural Econometrics for Psychologists." *Journal of Economic Psychology* 31: 553–76.

Angrist, J., and J.-S. Pischke. 2009. *Mostly Harmless Econometrics.* Princeton, NJ: Princeton University Press.

Banerjee, A., and E. Duflo. 2011. *Poor Economics.* New York: Public Affairs.

Blaug, M. 1980. *The Methodology of Economics.* Cambridge: Cambridge University Press.

Cartwright, N. 1980. "The Reality of Causes in a World of Instrumental Laws." In *PSA 1980*, vol. 2, edited by P. Asquith and R. Giere, 38–48. East Lansing, MI: Philosophy of Science Association.

———. 1983. *How the Laws of Physics Lie.* Oxford: Oxford University Press.

Coyle, D. 2007. *The Soulful Science.* Princeton, NJ: Princeton University Press.

Dennett, D. 1987. "Evolution, Error, and Intentionality." In *The Intentional Stance*, 287–321. Cambridge, MA: MIT Press.

Deutsch, D. 2011. *The Beginning of Infinity.* London: Allen Lane.

Dupré, J. 2001. *Human Nature and the Limits of Science.* Oxford: Oxford University Press.

Friedman, M. 1953. *Essays in Positive Economics.* Chicago: University of Chicago Press.

Hildenbrand, W. 1994. *Market Demand.* Princeton, NJ: Princeton University Press.

Kincaid, H. 2009. "Explaining Growth." In *Oxford Handbook of Philosophy of Economics*, edited by H. Kincaid and D. Ross, 455–75. Oxford: Oxford University Press.

Ladyman, J., and D. Ross. 2007. *Every Thing Must Go: Metaphysics Naturalized.* Oxford: Oxford University Press.

———. 2013. "The World in the Data." In *Scientific Metaphysics*, edited by D. Ross, J. Ladyman, and H. Kincaid, 108–50. Oxford: Oxford University Press.

Leamer, E. 2008. *Macroeconomic Patterns and Stories.* Berlin: Springer.

———. 2012. *The Craft of Economics.* Cambridge, MA: MIT Press.

Mirowski, P. 1989. *More Heat Than Light.* New York: Cambridge University Press.

Morgan, M. 2012. *The World in the Model.* Cambridge: Cambridge University Press.

Morgan, M., and M. Morrison. 1999. "Models as Mediating Instruments." In *Models as Mediators,* edited by M. Morgan and M. Morrison, 10–37. Cambridge: Cambridge University Press.

Papineau, D. 1978. *For Science in the Social Sciences.* Basingstoke, Houndmills: Palgrave Macmillan.

Rodrik, D. 2015. *Economics Rules.* New York: Norton.

Rosenberg, A. 2011. *The Atheist's Guide to Reality: Enjoying Life without Illusions.* New York: W. W. Norton.

Ross, D. 2005. *Economic Theory and Cognitive Science: Microexplanation.* Cambridge, MA: MIT Press.

———. 2010. "Should the Financial Crisis Inspire Normative Revision?" *Journal of Economic Methodology* 17: 399–418.

———. 2014. *Philosophy of Economics.* Basingstoke, Houndmills: Palgrave Macmillan.

Samuelson, P. (1947) 1983. *Foundations of Economic Analysis.* Enlarged ed. Cambridge, MA: Harvard University Press.

Smolin, L. 2006. *The Trouble With Physics: The Rise of String Theory, the Fall of a Science, and What Comes Next.* New York: Houghton Mifflin Harcourt.

Sutton, J. 2000. *Marshall's Tendencies.* Cambridge, MA: MIT Press.

———. 2001. *Technology and Market Structure.* 2nd ed. Cambridge, MA: MIT Press.

Van Fraassen, B. C. (2002). *The Empirical Stance.* New Haven, CT: Yale University Press.

Wallace, D. 2012. *The Emergent Multiverse: Quantum Theory according to the Everett Interpretation.* Oxford: Oxford University Press.

Winch, P. 1958. *The Idea of a Social Science and Its Relation to Philosophy.* London: Routledge and Kegan Paul.

Woodward, J. 2003. *Making Things Happen.* Oxford: Oxford University Press.

13

Why Really Good Science Doesn't Have All the Answers

MICHAEL RUSE

I am a historian and philosopher of science. I love science, and I can truly say that trying to understand it has led to a life of great satisfaction. I have never wanted to be a scientist—I have other aims as a scholar—but I can say that I respect it and its practitioners hugely. For me, it is one of the great achievements of humankind. I think a breakthrough like Charles Darwin's theory of evolution through natural selection is up there with *Così fan tutte* or Vermeer's *Christ in the House of Mary and Martha* or *Bleak House* or the poetry of Emily Dickinson. I can think of no higher praise.

This said, those who would push science beyond its limits—those infatuated or seduced by scientism, the belief that there is nothing beyond the grasp of science—do it no favor at all. I cannot give birth to a child. Does this make me any less a human being? An evening of Mozart does not fill an empty stomach. Does this make Mozart any less a composer or *Così* less of an opera? Jesus would not make bread from a stone. Does this make him any less the Messiah? To say that science has limits detracts not at all from its glory. To claim that science does have limits is to respect what it can do and do very well indeed.

In this chapter, I want to show that by its very nature—by the very things that make it so successful—science is bound to have limits, and I want to show also what some of those limits might be. Like William Whewell (1840), I am somewhat of a historicist about my philosophy of science. I recognize that some of the examples of where I suggest science cannot go may in the future be conquered by science. But I know that there will always be some areas or topics that cannot be conquered, and I have a good sense that we can identify some of them with confidence.

One more point before I begin. I am always suspicious of totally new approaches or totally new arguments. As Whitehead said wisely, they are either worthless or they are in Plato. I am not quite sure how far back you can trace what I have to say. Perhaps there are intimations in Plato's dialogues. I do know that what I am using to make my case is fairly standard thinking about the nature of science. I am going to rely on relatively noncontroversial points, drawn from both the history and the philosophy of science, and that makes me comfortable. To start my argument, let me make two major claims, one drawn from the philosophy of science and the other from the history of science.

Metaphor

A lot of people think that the aim of science is to tell it like it is. In the phrase made famous by Joe Friday from *Dragnet*: "Just the facts, ma'am, just the facts." As soon as you start to look seriously at science, that illusion vanishes quickly. Science is more than the facts, and it wants to be and needs to be more than the facts. Think of the aim of science. It is in some sense to tell us about the empirical world, the world of experience—going back to someone like John Locke (1689), the world of sensation and reflection. What is meant by "tell us about"? Well, obviously description. We want to know what the planets are doing. Are they going in circles or in squares or what? But more than that. We want some level of understanding, of explanation. Why do the planets go (as they do) in ellipses with the Sun at one focus, rather than in circles or squares? Without digging too deeply, such understanding is going to be based on the fact that experience is not a random affair but seems to be a function of regularity, of obedience to general rules or laws.

As is well known, about fifty years ago philosophers of science like Carl Hempel (1965; 1966) made significant progress in articulating these insights through such notions as the "covering law model of scientific explanation." But, although there was a lot of debate back then over these issues, everyone saw that there was more to science than just understanding. In some sense we want it to be a vital thing, moving forward and dragging or pushing us forward with it. We need science to make predictions. Don't just tell us about what happened in the past, but tell us about what will happen in the future. There was an eclipse of the Sun last week. When's the next one going to occur? I crossed these two white plants and got a red offspring. Don't just tell me why that happened, but tell me how to do it again. Can I get yellow or blue if I play my cards right? This Catholic priest has spent the last ten years feeling up little boys. Will this therapy, is there any therapy able to, change him so he is safe with small children?

So we want prediction, and again, people like Hempel did sterling work in showing just how prediction occurs and what the connections are between explanation and prediction. But as everyone realized, we need more than just prediction. We got a red flower, when will we get another? We need what is sometimes called "predictive fertility," where the emphasis is on the fertility (McMullin 1983). A good scientific theory is forward looking, pushing us to new insights and in new directions. It isn't just saying where it is, but it is going out into new territory. Are there features other than color that we can predict, and is all this confined to plants or is there a message here for those who work with animals? This is where the creationists are so wrong when they complain about the unsolved problems of science. Really good science generates problems, things to get the scientist fired up and working in unexplored fields. Good science has a strongly heuristic side to it.

What is the key to predictive fertility? What makes a theory forward looking in this way? In a word, metaphor! Everyone knows that science is deeply and thoroughly metaphoric, in the sense of using metaphors over and over again (Hesse 1966; Lakoff and Johnson 1980; Kuhn 1993). And this applies to all areas of science and not just the soft underbelly like sociology. Force, work, pressure, attraction, repulsion, big bang, black hole, pump, code, gene pool, drift, natural selection, tree of life, Oedipus complex—and so the story goes. There are people who think this a weakness, a bit as Plato regarded the use of diagrams by geometers. We should get rid of them as soon as possible. In the words of Thomas Hobbes (1651), "In reckoning, and seeking of truth, such speeches are not to be admitted." However, general opinion now is that metaphors are here to stay, and a good thing, too. Take "natural selection," the major metaphor from evolutionary biology, my own field of interest. Natural selection was controversial from the start, with many arguing that Darwin was illicitly bringing in a deity to do the selecting, analogously to the farmer or fancier who does the selecting in the animal and plant world. And there is reason to think this. Particularly in the early versions of Darwin's theory, the deity seems at work:

Let us now suppose a Being with penetration sufficient to perceive differences in the outer and innermost organization quite imperceptible to man, and with forethought extending over future centuries to watch with unerring care and select for any object the offspring of an organism produced under the foregoing circumstances; I can see no conceivable reason why he could not form a new race (or several were he to separate the stock of the original organism and work on several islands) adapted to new ends. As we assume his discrimination, and his forethought, and his steadiness of object, to be incomparably greater that those qualities in man, so we may suppose the beauty and complications of the adaptations of the new races and their differences from the original stock to be greater than in the domestic races produced by man's agency: the ground-work of his labours we may aid by supposing that the external conditions of the volcanic island, from its continued emergence and the occasional introduction of new immigrants, vary; and thus to act on the reproductive system of the organism, on which he is at work, and so keep its organization somewhat plastic. With time enough, such a Being might rationally (without some unknown law opposed him) aim at almost any result. (Darwin 1909, 80)

However, while there were those who actually thought that bringing in a deity was no bad thing, Darwin rushed to say that he intended no such thing:

Several writers have misapprehended or objected to the term Natural Selection. Some have even imagined that natural selection induces variability, whereas it implies only the preservation of such variations as occur and are beneficial to the being under its conditions of life. No one objects to agriculturists speaking of the potent effects of man's selection; and in this case the individual differences given by nature, which man for some object selects, must of necessity first occur. Others have objected that the term selection implies conscious choice in the animals which become modified; and it has even been urged that as plants have no volition, natural selection is not applicable to them! In the literal sense of the word, no doubt, natural selection is a misnomer; but who ever objected to chemists speaking of the elective affinities of the various elements?—and yet an acid cannot strictly be said to elect the base with which it will in preference combine. It has been said that I speak of natural selection as an active power or Deity; but who objects to an author speaking of the attraction of gravity as ruling the movements of the planets? Every one knows what is meant and is implied by such metaphorical expressions; and they are almost necessary for brevity. So again it is difficult to avoid personifying the word Nature; but I mean by Nature, only the aggregate action and product of many natural laws, and by laws the sequence of events as ascertained by us. With a little familiarity such superficial objections will be forgotten. (Darwin 1861, 84–85)

What is this powerful thing about natural selection that makes some people uncomfortable even to this day? It is that it demands that we look upon the living world in terms of *design*. We are to regard animals and plants as if they had been fashioned by a Creator, for the ends of success in the struggle for existence. Just as the farmer selects his fattest pigs or shaggiest sheep to achieve his ends of even-fatter pigs and even-shaggier sheep, or the breeder selects the bravest bulldogs or the fancier the finest songbirds, so the Creator selects the best-camouflaged butterflies or the hardiest oak trees or the nigh-indestructible parasite or the cleverest monkey or the fastest antelope. And notice how this metaphor works both from the viewpoint of explanation and from the viewpoint of prediction. It is very fertile, incredibly heuristic. Think of an area—vertebrate paleontology—where one really does need such a tool. People dig up the massive bones of brutes long gone, and try to make sense of them. No one today has seen anything quite like them, so how do we proceed? Use as an example the massive plates that went all the way down the back of the dinosaur *Stegosaurus*. Now, without some kind of guide, we are stuck. There are the plates, the beasts are extinct, and we are literally speechless. But with the metaphor of design, we can start to move forward.

To begin, things of that nature and size and intricacy don't just happen. Archdeacon Paley knew the score:

In crossing a heath, suppose I pitched my foot against a *stone*, and were asked how the stone came to be there; I might possibly answer, that, for any thing I knew to the contrary, it had lain there for ever: nor would it perhaps be very easy to show the absurdity of this answer. But suppose I had found a *watch* upon the ground, and it should be inquired how the watch happened to be in that place; I should hardly think of the answer which I had before given, that, for any thing I knew, the watch might have always been there. (Paley [1802] (1819), 1)

Why is this?

For this reason, and for no other, viz. that, when we come to inspect the watch, we perceive (what we could not discover in the stone) that its several parts are framed and put together for a purpose, *e.g.* that they are so formed and adjusted as to produce motion, and that motion so regulated as to point out the hour of the day; that, if the different parts had been differently shaped from what they are, of a different size from what they are, or placed after any other manner, or in any other order, than that in which they are placed, either no motion at all would have been carried on in the machine, or none which would have answered the use that is now served by it. (Ibid.)

So when we look at the plates on the *Stegosaurus*, we are looking for purpose. In this context, we are asking what end the plates might serve to make the dinosaur better at surviving and reproducing.

And a number of solutions have been proffered, and in good Popperian fashion—bold conjectures, rigorous refutations—most of them have been knocked down by the physical evidence. Could they be weapons of attack or defense? Not very likely, because they are pretty fragile. They just wouldn't be very good in a fight, unlike (say) the antlers on a deer. Could they be there for sexual attraction? Again, not very likely, because they are possessed by both males and females. Only the peacock has the tail feathers. The peahen is very drab. Could they be there to regulate the temperature of the cold-blooded animal, catching Sun's rays in the morning and cooled by the breezes in the heat of the day? Much more likely, because it does seem that they are "designed" to carry maximum amounts of blood and to catch heat and cool in breezes (Farlow, Thompson, and Rosner 1976). The closest analogy—and it is close—is the kind of plate you find in electricity generating stations where the aim is to transfer heat quickly and efficiently.

The question is not whether this is the end of the discussion. Longtime experience with science makes us very skeptical of inquiry-ending solutions. But we do have something—something with good evidence—to go on, where before there was just ignorance. Thanks to the metaphor of natural selection we are ahead. And note we are ahead because the metaphor suggests design. If we tried to strip the metaphor out of the discussion, speaking for instance of just something like "differential reproduction," we simply would have had no way to probe into the use of the plates and why they exist all down the back of the *Stegosaurus*.

In an analogous context, the British philosopher Richard Braithwaite (1953) said that the "price of the use of models is eternal vigilance." This is true in the case of natural selection. Perhaps, as the late Stephen Jay Gould used to insist, we see far too many instances of design where basically there is just chance and randomness (Gould and Lewontin 1979). Perhaps, for instance, things like male nipples just have no good reason, and striving to find such reasons is a case where our metaphor is time wasting. But eternal vigilance is one thing. Rejection or elimination is another. We need metaphors in science, and that is an end to it.

Or not quite, because there is another point to be stressed. One of the strengths of metaphors is that they focus you on a problem. They not only direct you forward, but if you listen to them they tell you where not to waste your energy. Take a metaphor like "my love is a rose." At once, you start to think of my girlfriend in a particular way. She is beautiful. She is fresh. She is not stinky like a cooked brussels sprout. If you are a bit of a joker, you might be hinting that she can be difficult. She is prickly like a rose. But what you are not saying is whether she is an evangelical or an atheist. You are not saying if she is a maths major or cannot put two and two together and always get the same answer. The metaphor doesn't talk about this and basically tells you not to waste your time on this. The same in science—as was pointed out by Thomas Kuhn (1962), who came to identify his famous notion of a paradigm with metaphor in some sense. In thinking of things produced by natural selection, keep within the bounds of design. In thinking about things on this Earth, for instance, don't spend your time worrying about relativity theory. Black holes are all very well, but not that relevant to the plates on the back of the *Stegosaurus*. To use a metaphor about metaphors, metaphors put blinkers on your inquiry. They are powerful at least in part because they rule certain questions out of court. It does not mean that the questions are bad questions or unanswerable—I can tell you if my girlfriend is an evangelical or atheist— but that they are not what you are talking about under the metaphor in use.

From Organicism to Mechanism

Turn now to the history of science. One thing very obvious to students of metaphor is that not all metaphors are equal. Some, as it were, include others (Lakoff and Johnson 1980). Take the metaphor of being upright as being good or healthy in some sense. It is easy to see where it comes from, at least in the case of bipedal organisms like humans. And it is as easy to see that it encompasses many subsidiary metaphors. "Stand up, stand up, for Jesus." "I'm a bit down in the mouth today." "You snake!" (This one may well owe something to

Genesis as well.) "He is a fine, upstanding young man." "I have never felt quite so low." "She showed real backbone in that issue." "It gives me the creeps." "A man never stands so tall as when he kneels to help a child." I can forgive the Shriners a lot of silly hats for that one.

These encompassing metaphors are known as "root" metaphors. In evolutionary biology, "design" is a root metaphor, as you explore the different kinds of design that one might have, and indeed as you look at the limitations of design. Overall in the history of Western science, historians identify two major root metaphors. The first, going back to the Greeks, is of the world as an organism, or at least of the world and its parts to be understood in terms of the organic (Ruse 2013). There is no reason why anyone must think of the seasons as lifelike, but people did—and obviously some still do. One thinks of the Earth as having a living rhythm—spring, summer, harvest, and then old age and finally winter and death in some sense. Plato was explicit in the *Timaeus*, seeing the whole universe as a living being. Of the Demiurge that fashioned the universe, Plato wrote:

> The god reasoned and concluded that in the realm of things that are naturally visible no unintelligent thing could be as a whole better than anything which does possess intelligence as a whole, and he further concluded that it is impossible for anything to come to possess intelligence apart from soul. Guided by this reasoning he put intelligence in soul, and soul in body and so he constructed the universe. (*Timaeus* 30b–c, in Cooper 1997)

Aristotle did not go the whole hog, as it were, and think of the universe—or the Earth—as a living being as such, but he certainly considered it appropriate to think of the world in organic terms, and this included the inorganic (Sedley 2008). Famously, he distinguished four kinds of cause. Taking them in turn, we start with "material" cause, the stuff from which something is made: the marble or bronze of a statue. Next comes "formal" cause, the pattern or archetype behind the creation: *The Thinker* has behind it the idea of a man deep in thought and oblivious to all else. The "efficient" cause is the maker: Rodin. And finally we have the "final" cause, the end or purpose for the creation: at one level it was a commission, but we can say that in an important sense it is to capture the essence of human nature, that which sets us little lower than the angels. Final cause is obviously in some sense organic, in that we are thinking of ends, which is just what we get with organisms. As it happens, Aristotle does not postulate a designer in the way of Plato. Rather, there seems to be something of the nature of things themselves—internal to things—that involves organization or intention or purpose. This is not necessarily intelligent, although of course it can be, but rather a kind of vital force that leads to organization. Note that for Aristotle, none of this is theological, as it would be later for Paley. He thought that this is the way that the world is, and hence it is appropriate to use these categories in one's science.

All this came to an end with the Scientific Revolution, the period in the sixteenth and seventeenth centuries roughly from the publication of Copernicus's work on the heavens in 1543 to the publication of Newton's work on mechanics in 1687. It was not a reaction against Christianity. Anything but:

> No Christian could ultimately escape the implications of the fact that Aristotle's cosmos knew no Jehovah. Christianity taught him to see it as a divine artifact, rather than as a self-contained organism. The universe was subject to God's laws; its regularities and harmonies were divinely planned, its uniformity was a result of providential design. The ultimate mystery resided in God rather than in Nature, which could thus, by successive steps, be seen not as a self-sufficient Whole, but as a divinely organized machine in which was transacted the unique drama of the Fall and Redemption. If an omnipresent God was all spirit, it was the more easy to think of the physical universe as all matter; the intelligences, spirits and Forms of Aristotle were first debased, and then abandoned as unnecessary in a universe which contained nothing but God, human souls and matter. (Hall 1954, xvi–xvii)

From thinking of the world as an organism, or at least in organic terms, the new root metaphor was that of the machine—the clock in Strasbourg Cathedral was a favorite. The world was nothing but dead matter subject to never-changing, never-ceasing laws. Obviously, certain kinds of causes were going to make the transition. Material and efficient causes particularly, usually rolled into one. Formal causes were often thought of as part of final causes—if I have an idea of the statue, this is very much bound up with my intention—although if one were so inclined, formal causes could be rolled in with material and efficient causes. But what about final causes? These really are different. You might think that they would make the transition readily. After all, you can ask about the point or endpoint of a machine. The automobile is built with transportation in mind. The guillotine with head-severing in mind. However, rapidly it became apparent that in the physical sciences, at least, talk of final causes really was not very helpful. It did not lead to new insights. Francis

Bacon thought of them as Vestal Virgins, decorative but useless. In the words of one of the very great historians of the Scientific Revolution, God became a "retired engineer" (Dijksterhuis 1961). His purposes were confined to theology and banished from science.

It is true that for the machine metaphor—mechanism—to become all triumphant, something had to be done about biology. It was clear that animals and plants do have purposes, or at least their parts do. It took Charles Darwin and natural selection to solve that conundrum. Although, as we have seen, the design metaphor is still used, as we have also seen ontologically it means nothing. We use the metaphor to ferret out purposes, but ultimately it is all molecules in motion without further meaning. Richard Dawkins has remarked on this speaking of organisms (including us humans) as "survival machines":

> Different sorts of survival machines appear very varied on the outside and in their internal organs. An octopus is nothing like a mouse, and both are quite different from an oak tree. Yet in their fundamental chemistry they are rather uniform, and, in particular, the replicators which they bear, the genes, are basically the same kind of molecule in all of us—from bacteria to elephants. We are all survival machines for the same kind of replicator—molecules called DNA—but there are many different ways of making a living in the world, and the replicators have built a vast range of machines to exploit them. (Dawkins 1976, 22)

More recently than Darwin, many feel that the final unknown territory has been entered and conquered. Now even the human brain and the associated thought have been brought beneath the root metaphor. The philosopher Andy Clark writes:

> The computer scientist, Marvin Minsky once described the human brain as a meat machine—no more, no less. It is, to be sure, an ugly phrase. But it is also a striking image, a compact expression of both the genuine scientific excitement and the rather gung-ho materialism that has tended to characterize the early years of cognitive science research. Mindware—our thoughts, feelings, hopes, fears, beliefs, and intellect—is cast as nothing but the operation of the biological brain, the meat machine in our head.

He continues:

What exactly is meant by casting the brain as a machine, albeit one made out of meat? There exists a historical trend, to be sure, of trying to understand the workings of the brain by analogy with various currently fashionable technologies: the telegraph, the steam engine, and the telephone switchboard are all said to have had their day in the sun. But the "meat machine" phrase is intended, it should now be clear, to do more than hint at some rough analogy. For with regard to the very special class of machines known as computers, the claim is that the brain (and, by not unproblematic extension, the mind) actually *is* some such device. It is not that the brain is somehow *like* a computer: everything is like something else in some respect or other. It is that neural tissues, synapses, cell assemblies, and all the rest are just nature's rather wet and sticky way of building a hunk of honest-to-God computing machinery. Mindware, it is then claimed, is found "in" the brain in just the way that software is found "in" the computing system that is running it. (Clark 2000, 7–8)

Well, yes, but obviously we do still have a metaphor, even if one conceals it by extending the meaning of *machine*. Machines are not made out of meat. They are made out of plastic and copper and so forth. But in a way, this doesn't really matter for our ends. The point is that as we move into and through the twenty-first century the machine metaphor rides triumphant.

Unasked Questions

Now let me start to put things together. I stress again—and this is a very important part of my argument—in neither philosophy nor history am I saying anything outrageous or that would not be considered totally standard. Of course, as is the nature of things, there are lots of variations we can spin, and no doubt all over North America today there are graduate students writing dissertations showing that these claims are false. But grownups know otherwise. And the question to which my discussion is leading is, given that the machine metaphor is dominant in modern science, the root metaphor that structures the way science is done, what are the blinkers that come with it? In other words, what questions are simply not asked—and hence not answered—by modern science?

Let me suggest four such questions (Ruse 2010). First, there is what is known, thanks to Martin Heidegger (1959), as the "fundamental question of metaphysics": why is there something rather than nothing? Notice that whatever else, this is not a regular scientific question to which an answer like "the big bang" will suffice. It is more about the very nature of existence itself. If you take things back through a chain of efficient causes, you are either going to keep going infinitely or you are going to come to a stop with a first cause. But then you still have the question of why there is this infinite chain or why the first cause turned up at all. I want to suggest that at least part of the puzzle here—perhaps only a small part, because I am certainly not going to answer the fundamental question!—is that people don't see that the question is by its very nature not one that modern science could answer. With machines, it is a bit like Hannah Glasse's first catch your hare.[1] Suppose you are making an automobile. If you work for General Motors, I am sure you are going to be very concerned about supplies. Where is the steel to come from, and what about the plastics? But at some point you just give up looking backward. Ultimate origins are just not what machines are about. If suppliers for General Motors start worrying about the big bang, they are soon going to be out of business. In short, what I am arguing is that the machine metaphor is terrific, but it has to take things—ultimate substances—for granted and get on with the job. It cannot and should not do everything.

Second, morality. I am being totally unoriginal—actually, I am being in the tradition of David Hume—in pointing out that in the modern world, the world of machines, you don't get values from facts:

> In every system of morality, which I have hitherto met with, I have always remarked, that the author proceeds for some time in the ordinary ways of reasoning, and establishes the being of a God, or makes observations concerning human affairs; when all of a sudden I am surprised to find, that instead of the usual copulations of propositions, *is*, and *is not*, I meet with no proposition that is not connected with an *ought*, or an *ought not*. This change is imperceptible; but is however, of the last consequence. For as this *ought*, or *ought not*, expresses some new relation or affirmation, 'tis necessary that it should be observed and explained; and at the same time that a reason should be given, for what seems altogether inconceivable, how this new relation can be a deduction from others, which are entirely different from it. (Hume [1739] 1940, 302)

That's the point about machines—they are neither moral nor immoral, they are amoral. Is the guillotine a good thing? I don't think so, but most of my fellow Floridians are enthusiastically in favor of it. We read the values in rather than read the values out. Another area where modern science is silent.

Third, and now I start to get more controversial, there is the matter of consciousness, of sentience. What is known as the "hard question" (Chalmers 1996). Some people, Daniel Dennett (1992) is a notorious example, think that the problem is solved and that it can be given a purely mechanical solution. Talk about the workings of the brain and you have the answer to sentience. I will not dwell on this, because I share the general opinion: this is simply to declare the battle won. As was suggested for the conflict in Vietnam, say we have a victory and go home. Unfortunately, although this may have been good political advice, it won't do in the world of philosophy. Sentience is just something different, and as Leibniz pointed out, machines just don't hack it:

> One is obliged to admit that *perception* and what depends upon it is *inexplicable on mechanical principles*, that is, by figures and motions. In imagining that there is a machine whose construction would enable it to think, to sense, and to have perception, one could conceive it enlarged while retaining the same proportions, so that one could enter into it, just like into a windmill. Supposing this, one should, when visiting within it, find only parts pushing one another, and never anything by which to explain a perception. Thus it is in the simple substance, and not in the composite or in the machine, that one must look for perception. (Leibniz 1714, §17)

There are those, I am one, who think the problem will never be solved. We think we are no closer to a solution than we were back in the seventeenth century. Computers tell us a lot about the working of the brain. They don't even scratch the question of sentience. Not everyone agrees with us "new mysterians"—Paul and Pat Churchland, for example, think that although there is no solution yet, there may be one in the future (Churchland P. M. 1995; Churchland P. S. 1986). I cannot say absolutely that they are wrong—this is what I mean by being somewhat of a historicist like Whewell—but I am not holding my breath. I see no solution, and I see no possibility of one coming through modern science. Machines can calculate, but they just don't think.

Fourth and finally, I would add questions about the purpose of it all. Admittedly, machines do have purposes, but we have seen that modern science filters out this aspect of the metaphor. The Nobel Laureate in physics Steven Weinberg has said, "The more the universe seems comprehensible, the more it also seems pointless" (Weinberg 1977, chap. 8, p. 322). Why am I not surprised? Modern science does not set out to answer this kind of question, so you are going to be out of luck if you ask it. To quote Richard Dawkins again, "In a universe of blind physical forces and genetic replication, some people are going to get hurt, other people are going to get lucky, and you won't find any rhyme or reason in it, nor any justice. The universe we observe has precisely the properties we should expect if there is, at bottom, no design, no purpose, no evil and no good, nothing but blind, pitiless indifference" (Dawkins 1995, 133).

Christianity by the Back Door?

So this in essence is why I am not into scientism. I think that thanks to the machine metaphor, there are going to be questions that science does not even set out to answer let alone actually answer. I have given four suggestions. There may be more. Perhaps the free will problem is a candidate. Some may get answered or closed off. I don't know. I do know that there will always be some, and I would be very surprised if none of my candidates persisted. But before I finish, there is a question that, although strictly I do not have to answer it, morally I must speak to it. What, then, do I think is the status of my four questions? Do I think they have nonscientific answers? And if so, what is the nature of these nonscientific answers? Specifically, am I pursuing a Christian agenda by posing questions to which the Christian has ready answers? Am I saying that science is not enough and that we need religion?

Well, that is a question with an answer. Absolutely not! It is true that the religious, the Christian specifically, has answers to these questions. Why is there something rather than nothing? Because a good God created freely out of nothing. What is the foundation of morality? The will of God. What is sentience? Being made in the image of God. What is the point of it all? To love and obey and worship God, hoping to enjoy eternal bliss in his company. But there is no reason why, even if science cannot speak to these issues, philosophy and theology should not step in here and show the inadequacies of the Christian answers. The Christian—or any religious person—has the right to offer answers. Such individuals are not in conflict with science as they are when claims are being made about the empirical world, for instance that there was a universal flood shortly after the creation of the Earth. But it does not mean that we are compelled to accept the answers.

First, the good creator God. He has to be a necessarily existing being;

otherwise, he is part of the efficient causal chain like everything else. God in some sense has to be the ground of our being, and he himself must be outside time and space and eternal. I am not saying anything that is not standard theology. But as philosophers have pointed out, Hume and Kant specifically, the very idea of a necessarily existing being is, to say the least, highly problematic:

> We're faced with a strange and bewildering fact, namely, that while the inference from "Something exists" to "An utterly necessary being exists" seems to be compelling and correct, when we try to form a concept of such a necessity—i.e. a concept of something's necessarily existing—we find that we can't overcome the obstacles that the understanding puts in our way through its requirements for what such a concept would have to be like. (Kant [1781] 1929, 568)

I don't have to go into the issue in depth here. It is enough to show that foregoing scientism does not push someone into the hands of the Christian God.

Much the same sorts of comments apply to the other three questions. As it happens, I am one who thinks that Darwinian evolutionary theory makes improbable an objective basis for ethics—I am what is known in the trade as a "debunker." But I don't get this conclusion from science alone but from philosophy also. So if you want to argue against me it is quite possible to do so, just as if you want to argue against the Christian's belief that morality lies in the will of God—perhaps by invoking the *Euthyphro* argument (does "good" mean what God wants or does God want what is good?)—you are free to do so. The point is that science does not provide the answer to an objective ethics, but it doesn't mean that discussion stops or that someone is pitchforked into taking a religious position.

In the case of sentience, I find the Christian answer about as unsatisfying as that of Dennett. Indeed, I am inclined to think it worse. To say that we are made in the image of God is but the start of the discussion, and unpacking what this might mean is a formidable task. God is unchanging, eternal, outside space and time (Ruse 2015). What kind of thought processes can he have? How can I, who changes his mind about every ten seconds, be meaningfully made in the image of God? It is generally thought, following people like Augustine, that rationality is the key here. That would fit with the saint's Neoplatonic belief that the rational part of the soul should control the other parts. But how exactly that translates in the case of God is a mystery—which may be satisfactory to the believer, but is not much help to the rest of us.

And finally, turning to ultimate purposes, there are massive philosophical difficulties associated with life after death, especially if you adopt the Christian position that nothing much is going to happen until the Day of Judgment. Why should you say that the Michael Ruse who is writing this essay is the same Michael Ruse enjoying eternal bliss with the Creator, or—perhaps more likely—facing an infinite number of infinitely big piles of coals, all of which have to be shoveled into an infinite number of fiery furnaces? I have seen the suggestion that God keeps our souls between death and resurrection in his collection of divine CDs (Polkinghorne 1994). We are the software that he plays in the due course of time. Perhaps so, although since apparently God has made one copy of my soul, what is to stop him from making two or three copies or, since you can never have too much of a good thing, an infinite number of copies? Enough of such happy speculations. The point is made that although science may say nothing about the hereafter, it does not mean we are powerless to counter the Christians who made grandiose claims on the basis of his or her faith.

So what are we left with? There are those, like Wittgenstein (1965), who argue that at least some of these questions can never be answered and so are meaningless. Again, it seems to me that this is a Dennett-type solution. I have no answer to the fundamental question, but I don't think it is a silly or meaningless question. It is a puzzle why there is existence, and rejecting God does not make it less of a puzzle. At best—although I am not sure that this is such a bad thing—I think we have to be skeptics or agnostics. We just have no answers. Perhaps paradoxically, modern science comes to our aid here. Is the electron a wave or a particle? It cannot be both, and yet it exhibits features of both. Thanks to Heisenberg's uncertainty principle we just cannot say. We have no answer or prospect of an answer. It's the same with the questions unanswered by science. We have no answer or prospect of an answer. Which leads me to conclude with yet another reflection of Richard Dawkins: "Modern physics teaches us that there is more to truth than meets the eye; or than meets the all too limited human mind, evolved as it was to cope with medium-sized objects moving at medium speeds through medium distances in Africa" (Dawkins 2003, 19). Amen.

••••••

Note to Chapter Thirteen

1. Actually, Joe Friday did not say "Just the facts, ma'am, just the facts," and Hannah Glasse did not say "first catch your hare"—but they could have. Friday's famous saying came from a parody, and Glasse wrote, "Take your hare when it be cas'd [skinned]."

Literature Cited

Braithwaite, R. B. 1953. *Scientific Explanation*. Cambridge: Cambridge University Press.

Chalmers, D. J. 1996. *The Conscious Mind*. New York: Oxford University Press.

Churchland, P. M. 1995. *The Engine of Reason, the Seat of the Soul*. Cambridge, MA: MIT Press.

Churchland, P. S. 1986. *Neurophilosophy: Towards a Unified Science of the Mind and Brain*. Cambridge, MA: MIT Press.

Clark, A. 2000. *Mindware: An Introduction to the Philosophy of Cognitive Science*. New York: Oxford University Press.

Cooper, J. M., ed. 1997. *Plato: Complete Works*. Indianapolis: Hackett.

Darwin, C. 1861. *Origin of Species*. 3rd ed. London: John Murray.

———. 1909. *The Foundations of the Origin of Species: Two Essays Written in 1842 and 1844*. Edited by F. Darwin. Cambridge: Cambridge University Press.

Dawkins, R. 1976. *The Selfish Gene*. Oxford: Oxford University Press.

———. 1995. *A River Out of Eden*. New York: Basic Books.

———. 2003. *A Devil's Chaplain: Reflections on Hope, Lies, Science and Love*. Boston: Houghton Mifflin.

Dennett, D. C. 1992. *Consciousness Explained*. New York: Pantheon.

Dijksterhuis, E. J. 1961. *The Mechanization of the World Picture*. Oxford: Oxford University Press.

Farlow, J. O., C. V. Thompson, and D. E. Rosner. 1976. "Plates of the Dinosaur Stegosaurus: Forced Convection Heat Loss Fins?" *Science* 192: 1123–25.

Gould, S. J., and R. C. Lewontin. 1979. "The Spandrels of San Marco and the Panglossian Paradigm: A Critique of the Adaptationist Programme." *Proceedings of the Royal Society of London, Series B: Biological Sciences* 205: 581–98.

Hall, A. R. 1954. *The Scientific Revolution 1500–1800: The Formation of the Modern Scientific Attitude*. London: Longman, Green.

Heidegger, M. 1959. *An Introduction to Metaphysics*. New Haven, CT: Yale University Press.

Hempel, C. G. 1965. *Aspects of Scientific Explanation*. New York: Free Press.

———. 1966. *Philosophy of Natural Science*. Englewood Cliffs, NJ: Prentice-Hall.

Hesse, M. 1966. *Models and Analogies in Science*. Notre Dame, IN: University of Notre Dame Press.

Hobbes, T. (1651) 1982. *Leviathan*. Harmondsworth, Middlesex: Penguin.

Hume, D. (1739) 1940. *A Treatise of Human Nature*. Edited by D. F. Norton and M. J. Norton. Oxford: Oxford University Press.

Kant, I. (1781) 1929. *Critique of Pure Reason*. Translated by N. Kemp Smith. New York: Humanities Press.

Kuhn, T. 1962. *The Structure of Scientific Revolutions*. Chicago: University of Chicago Press.

———. 1993. "Metaphor in Science." In *Metaphor and Thought*, 2nd ed., edited by Andrew Ortony, 533–42. Cambridge: Cambridge University Press.

Lakoff, G., and M. Johnson. 1980. *Metaphors We Live By*. Chicago: University of Chicago Press.

Leibniz, G. F. W. 1714. *Monadology and Other Philosophical Essays*. New York: Bobbs-Merrill.

Locke, J. 1689. *An Essay concerning Human Understanding*. Edited by A. C. Fraser. New York: Dover.

McMullin, E. 1983. "Values in Science." In *PSA 1982*, edited by P. D. Asquith and T. Nickles, 3–28. East Lansing, MI: Philosophy of Science Association.

Paley, W. (1802) 1819. *Natural Theology*. In *Collected Works*, vol. 4. London: Rivington.

Polkinghorne, J. C. 1994. *Science and Christian Belief: Theological Reflections of a Bottom-Up Thinker*. London: SPCK.

Ruse, M. 2010. *Science and Spirituality: Making Room for Faith in the Age of Science*. Cambridge: Cambridge University Press.

———. 2013. *The Gaia Hypothesis: Science on a Pagan Planet*. Chicago: University of Chicago Press.

———. 2015. *Atheism: What Everyone Needs to Know*. Oxford: Oxford University Press.

Sedley, D. 2008. *Creationism and Its Critics in Antiquity*. Berkeley: University of California Press.

Weinberg, S. 1977. *The First Three Minutes: A Modern View of the Origin of the Universe*. New York: Basic Books.

Whewell, W. 1840. *The Philosophy of the Inductive Sciences*. London: Parker.

Wittgenstein, L. 1965. "A Lecture on Ethics." *Philosophical Review* 74: 3–12.

14

Scientism (and Other Problems) in Experimental Philosophy

TOM SORELL

Conceptual analysis aims to uncover the system in our unreflective use of such philosophically important terms as *knows*, *causes*, and *appears*. At one time, this kind of investigation was aided by a catalogue of senses derived from past and present written usage in the *Oxford English Dictionary* (Austin 1969). But a more familiar form of analytic technique consists of proposing necessary and sufficient conditions for the truth of a schematized form of sentence containing a term under investigation. Philosophers ask what it is for *A* to know that *p*, or under what conditions it is true that *e* causes *e'*, or what makes it true that "A did *a* intentionally." Then they run through various previously proposed necessary and sufficient conditions, rejecting in turn any that are open to counterexample. What counts as a counterexample is a situation in which the proposed necessary and sufficient conditions obtain but the philosopher is disinclined to say, or is inclined to deny, that *A* knows that *p* or that *e* causes *e'* and so on. In these cases, the inclination or disinclination is driven by the philosopher's own (semantic) intuitions.

One strand of experimental philosophy seems to be defined by its *rejection* of the kind of appeal to intuitions built into this form of conceptual analysis.[1] Experimental philosophers complain that *knows* and *causes* are items of the *public* property of a linguistic community, and yet only a small number of language users—philosophers in their armchairs, or participants in philosophy classes or seminars, or contributors to philosophy journals—provide the linguistic intuitions that conceptual analyses are tested against (Knobe 2007). Experimental philosophers widen the sample of respondents beyond the usual set of interlocutors. Using their philosophical education as well as the methods of empirical psychology, they devise questionnaires that bring out the variety of intuitive response in the wider linguistic community, a variety that sometimes reflects cultural differences (Weinberg, Nichols, and Stich 2001).

The surveys of the experimental philosophers occasionally indicate an apparently significant departure of folk intuition from the intuitions of those in professional philosophy or their students. For example, Nichols and Knobe have found that subjects presented with the difference between a determinist and an indeterminist universe are overwhelmingly inclined to say that ours is indeterminist, and also to say that in a determinist universe no one is responsible for their "actions." But if a piece of serious wrongdoing by someone in the determinist universe is described in great detail, respondents say (again overwhelmingly) that that agent *is* responsible (Nichols and Knobe 2007).

Experimental philosophers have also identified ways of setting up or ordering examples that affect folk intuitions in unexpected ways. One study of intuitions elicited from a sample of nonphilosophers purports to show that significant variation results from the way examples, including Keith Lehrer's TrueTemp example, are juxtaposed (Swain and Weinberg 2008).

The question to be pursued in this chapter is whether such findings are a valuable corrective to traditional philosophical appeals to intuition, or whether they are a distraction motivated in part by scientism. It is hard to answer this question in full generality, because there is more than one kind of appeal to intuitions in philosophy, and because, as already noted, the interventions of experimental philosophers are not uniform either. Indeed, experimental philosophy seems to contain a spectrum of positions from, at one extreme, support for analytic philosophy informed by surveys to, at the other extreme, rejection of analytic philosophy and its characteristic concerns. The more that experimental philosophy is a distinctive position—that is, hostile to intuitions and analysis—the more it *does* seem to me to be an outright departure from philosophy or a distraction in philosophy.

My criticisms will focus on two unattractive aspects of experimental philosophy. The first is the vagueness of its self-image; the second is its scientism. Either it defines itself so vaguely that there is no difference between being experimental and being empirically informed, or else it distinctively depends on adopting some elementary experimental techniques from psychology, in which case its findings are often disputable, and often not easy to recognize as philosophy. Then there is its occasionally scientistic tendency. Some experimental philosophers criticize the appeal to intuitions in conceptual analysis or in certain metaethical discussions as a repudiation of empirical techniques where empirical techniques can be of service. More generally, some experimental philosophers believe in the strong continuity of philosophy with empirical science, and look back nostalgically to an era when *philosophy* and *science* were interchangeable terms, and the research programs of philosophy were subject to less of a division of intellectual labor (Knobe 2007, 120). I shall suggest that nostalgic experimental philosophy along these lines is scientistic in a sense I once discussed at book length (Sorell 1991). That is, it objectionably treats natural science as the preferred body of results and methods for intellectual work of every kind.

I shall contrast the objectionable scientism of some experimental philosophy with what is unobjectionable: nonexperimental philosophy that is empirically informed. This kind of philosophy employs traditional analytic techniques and recognizes traditional philosophical problems, but tries to maintain consistency with the results of social and natural sciences, and it treats many empirical findings as relevant to philosophical inquiry.[2] In short, empirically informed philosophy can use a priori methods of conceptual analysis while also taking account of a posteriori scientific results. Those results may or may not include surveys of the intuitions of nonphilosophers, as appropriate.

I shall limit my discussion in three ways. First, I am interested in experimental philosophers who treat orthodox philosophical appeals to intuitions as pronouncements from the armchair—that is, as dogmatic, or unduly a priori, or both.[3] Second, I shall be concerned with appeals to intuition in *moral* philosophy as opposed to epistemology or the philosophy of causation. Critics of experimental philosophy have already questioned the effectiveness of experimental philosophy as an approach to certain questions in metaethics (Kauppinen 2007). I shall consider uses of intuitions in *normative* ethics.

In normative ethics, the method of reflective equilibrium licenses appeals to intuitions in some of the ways experimental philosophers find objectionable. I shall defend the use of reflective equilibrium, and I shall claim that whatever its drawbacks are from the point of view of experimental philosophy, the method of reflective equilibrium lends itself perfectly well to being empirically informed, and being empirically informed is the antidote to the bad kind of philosophizing from the armchair. There are several reasons for being empirically informed in normative ethics that diverge from reasons for being empirically informed in metaethics. The reasons are to do with the practitioner-facing character of some normative ethics, and the fact that such ethics has to be informed by the context and training and preoccupations of the relevant practitioners or else lack credibility with its audience. Suitably informed normative ethics is empirically informed—informed about the topics the practitioners consider, and about their attitudes toward ethicists. But these reasons do not support an experimental approach to the deliverances of reflective equilibrium or, more generally, to moral philosophy. I shall conclude, then, that in normative moral philosophy the experimental turn is more distracting than instructive.

The rest of this chapter falls into three sections. In the first, I discuss some of the issues raised by the experimental philosophers' repudiation of armchair-based analytic philosophy. Then I consider briefly and reject the claim that, historically speaking, analytic philosophy is a sort of regrettable aberration: a departure from an empirically minded approach on the part of philosophers going back to antiquity—into which experimental philosophy fits very comfortably. Finally, I shall consider some good reasons for thinking that normative ethicists have to leave the armchair. Although I think these reasons are weighty, they do not support the experimental turn in philosophy.

The Repudiation of "Armchair Philosophy"

To distinguish empirically informed nonexperimental philosophy from experimental philosophy, it is important to consider their different ways of repudiating "armchair philosophy." Philosophers who are in favor of the empirically informed approach are willing to use scientific results not specifically addressed to philosophers when they seem relevant to philosophy. These results could range from case studies of psychopaths (relevant to philosophy of punishment or theories of rationality) to information about medical procedures (relevant to medical ethics) to empirical taxonomic methods (relevant to candidate terms for reductive identity statements in the semantics of natural kind expressions) (Beebee and Sabbarton-Leary 2010). Philosophers unwilling to consult this sort of empirical data might be called armchair philosophers in the pejorative sense of the phrase.

But a different sort of armchair philosopher is relevant to experimental philosophy. Experimental philosophers dislike *their* philosopher in the armchair because he or she appears to define philosophically important terms by reference to the semantic intuitions of the few rather than the many (Knobe 2007, 83). Experimental philosophers also object that intuitions are elicited in ways that disguise the psychological distortion and misinformation that affect most people. In armchair philosophy at its most extreme, only the intuitions of the armchair philosophers themselves are consulted, as if these philosophers had special semantic authority. And the intuitions of the armchair philosophers are not treated as if they, like the intuitions of the rest of humanity, were significantly affected by distorting influences.

But this picture of the practice of armchair philosophy is surely a caricature. In the era of ordinary language philosophy, the empirical results of lexicography were brought to bear on philosophy. The *OED* was in a way a representative of the linguistic community, and its entries—the result of empirical investigation about actual usage—constrained what could count as raw material for linguistic analysis.[4] So the image of philosophers, one by one, consulting only their *own* linguistic intuitions seems inappropriate even when applied to ordinary language philosophers.

What about practitioners of conceptual analysis after Gettier? Here the *OED* is no longer a prop for philosophical activity, and definition rather than gloss is attempted as a preliminary to an informative discussion of different kinds of knowledge conforming to a favored definition. The method for testing the favored definition is by attempts at self-refutation through counterexample. But it is all too easy to exaggerate the isolation of the philosopher and of the testing process involved in the method of counterexample. The medium for exposing would-be analyses to counterexample is not, as talk of the armchair suggests, a soliloquy. It is a sort of dialogue that unfolds in a philosophy class or seminar (Kauppinen 2007, 106), or, over a much longer period of time, in a journal article and its replies and sequels.

It is not as if armchair analysts need to take themselves to be incorrigible about the application of the relevant terms. They may even start to lose confidence, after devising a series of counterexamples, that their linguistic intuitions are reliable. This is where dialogue comes in. One of its functions is to reveal eccentricity in the application of terms, or overcomplication and undue artificiality in counterexamples. It can thus be of great interest in a class or in a reply to an article if people differ over what to say in a particular case, or do not know what to say. Certainly, it would be unusual for either disagreement or uncertainty simply to be disregarded. A relevant ambiguity or a linguistic eccentricity might have to be identified to clear up a disagreement, and a fresh example might have to be devised to get over linguistic uncertainty.[5]

The process of proposing definitions, then, does not consist of one philosopher talking to himself, still less one philosopher brooking no disagreement from fellow speakers of the language. Participants in a dialogue over a proposed analysis *each* appeal to their intuitions, and everyone is supposed to take note of, and try to explain, divergences. They exploit their abilities as native users of *knows* and *causes* and as people who are able after philosophical training to recognize what necessary and sufficient conditions are, and what an apparent counterexample is.

It might be thought that what is at issue is not the isolation of the individual philosopher from the wider linguistic community but the isolation of philosophers as a profession, or hosts of analytic philosophers wanting to capture linguistic intuitions, from the wider linguistic community. But here two responses may be in order. (1) There are well-known divergences in intuitions even between similarly trained philosophers. This suggests that broad similarity in social class, education, and gender is not sufficient to produce a kind of conceptual uniformity. Exposing semantic claims to audiences that are *more* diverse in ethnic, gender, and class terms, then, is not necessary to *introduce* diversity of intuition, since there never was a cozy unanimity in the first place. (2) Evidence from experimental philosophy of divergence between the analytic orthodoxy and the wider linguistic community is itself sometimes disputable. As Sosa points out with regard to the Knobe and Nichols experiment, the inclination to say that an agent is responsible for wrongdoing even in a determinist universe may be a matter of equivocation on *responsible*, which can mean either being *held* responsible or being responsible. And this is a general problem. Divergence in intuitions may arise from ambiguities in crucial terms, intuitive reactions to which are being elicited (Sosa 2006, 104).

There is also in the background of conceptual analysis an excerpt from the history of philosophy that gives, for example, the analysis of *knows* its point. The analysis is answerable not only to linguistic intuitions—however widely sampled—but also to a philosophical tradition in which it is a *problem* whether anyone knows anything. One constraint on the analysis of knowledge is that it should not *dissolve* the problem of skepticism but should, on the contrary, identify what it is about knowledge that might make it *seem* to some philosophers as if it is pretentious for anyone to claim to have it.[6] Many analyses do throw light on these questions, by proposing different norms of evidence for beliefs, some quite idealized and hard to satisfy, or different sources of unreliability in the mechanisms we use to form true beliefs. Both internalist and externalist analyses of knowledge can thus leave it open whether people actually know things, while identifying conditions that might make knowledge hard to acquire or easy to undermine.

It summarizes the preceding to say that far from being carried out in separation from a wider linguistic community at a time, or a philosophical community at a time and over time, analysis is both outward looking and temporally extended. This does not mean, necessarily, an openness to *scientific* results, but clearly there are examples from externalism in the theory of knowledge of precisely this. For instance, it is possible for a philosopher to hold a priori that knowledge is nonaccidentally true belief or undefeated justified true belief, and still think that experimental findings about gullibility are relevant to whether true beliefs formed by some subjects on the basis of testimony are accidentally true or undermined by defeating evidence (Coady 1992, 193–95). In other words, accidentally true belief or defeasibility conditions may have to be conceptualized with the help of psychology, cognitive and social.

Psychology, then, may perfectly well impinge on analysis. But this way of impinging is not the way of experimental philosophy: that is, it is not a wider than customary, and more systematic, sampling of relevant intuitions. Again, while externalist analyses, taxonomies of knowledge, and *theories* of kinds of knowledge can be open to the findings of empirical psychology, the impetus for the proposal that knowledge is nonaccidentally true belief can remain the traditional analytical one: the wish to correct defects in earlier would-be definitions of knowledge, defects revealed by counterexamples submitted to the intuitions of the usual closed circle of the philosophically initiated.

Nostalgia for the Prehistory of the Science/Philosophy Distinction

So far, we have been concentrating on semantic intuitions and philosophically important terms to describe armchair philosophy and what might be wrong with it. But the linguistic turn itself might be thought to be an aberration in philosophy. According to Knobe, the methods of experimental philosophers are more relevant than those of the analytic tradition to studying the mechanisms underlying various theoretical and practical capacities of human beings. What is more, still according to him, studying these mechanisms is more characteristic of the long-term history of philosophy, and thus might have a stronger claim to be authentic philosophical research, than work in the analytic tradition. Knobe concedes that experimental philosophy may not have lasting relevance for the "analytic project," except as a negative program intended to deflate some of the claims philosophers have based on intuition; but he suggests that questions about human nature and cognition and more generally the mechanisms underlying intentionality, cognition, and morality have been of interest in Western philosophy since Plato and Aristotle, and that philosophers as different from one another as Nietzsche and Hume have carried on this tradition.

Here is Knobe sketching the relevant line of thought:

[Experimental philosophers] focus on questions about the *internal psychological processes* underlying people's intuitions. That is to say, they are not primarily about the external properties and relations that people's concepts pick out but rather about the internal processes that lead people to have the intuitions that they do. By studying these processes, experimental philosophers take themselves to be getting at certain fundamental issues about the way people ordinarily understand their world. . . . Perhaps the claim [from analytic philosophers] is that research on the most fundamental concepts people use to understand themselves and their world doesn't really count as "philosophy." But this claim seems a bit hopeless and bizarre. It is not as though experimental philosophers are involved in some sort of radical departure from the traditional problems of philosophy. In fact the chronology is just the opposite. For most of the history of philosophy, questions about human nature and the nature of cognition were absolutely central. Then, for a comparatively brief period, many philosophers forsook these problems in favor of problems that had a more technical character. Experimental philosophy now seeks a return to the traditional problems of philosophy, the problems that played such a prominent role in the work of Plato, Aristotle and so many of their successors. (Knobe 2007, 89–91)

This argument from the history of philosophy is questionable on at least three grounds. First, and most obviously, it equivocates on the word *philosophy*. It is true that *philosophy* used to mean science in general, and included a whole range of systematic empirical inquiries, such as biology, physics, medicine, the study of the mind, morals, politics, and many other things. If Knobe's complaint were that philosophy more narrowly conceived has only recently broken away from the main body of science, and that there is a still a place for the empirical study of human nature and the nature of cognition, it would be uncontroversial. But what is at issue is not whether there is room for such empirical study but whether there is room for it as a branch of *philosophy*: it begs the question to argue from a description of the history of philosophy in which *philosophy* means "science." That description begs the question because the meaning of the term *philosophy* has changed.

The second ground for criticizing Knobe's argument is that it is not at all clear, as he claims, that experimental philosophy studies the processes that underlie intuitions. Many experimental philosophy papers seem to report the results of surveys of semantic intuitions elicited by examples similar to those that are used in philosophy classes. It is true that experimental philosophers offer hypotheses to explain the divergence of those intuitions from reactions that most philosophers expect.[7] But these hypotheses are very far from establishing which fine-grained cognitive or affective capacities generate the intuitions, still less which processes (presumably in the brain) realize the relevant capacities. In this respect, experimental philosophy is a far cry from the explanations that the neurosciences would give for distinctive losses of speech, writing, and reading skills associated with brain injury of different kinds. The neurosciences do indeed identify, or at least indicate, underlying processes or at least regions of the brain with certain cognitive or affective functions. Experimental philosophy, to judge by at least many of its most widely cited studies, does far less.

If this is right, the claim of experimental philosophy to belong to the future history of brain science might be weaker than its claim to fit into analytic philosophy today. I do not deny that experimental philosophy can very plausibly be claimed to be an unusual kind of *psychology*–psychology that fastens onto intuitions elicited by philosophically inspired examples as opposed to some more commonplace human behavior. But in that case, why insist on classifying it as philosophy, indeed as particularly authentic philosophy?

It is true that some people who call themselves experimental philosophers show less hostility than Knobe does to analytic philosophy. Far from thinking that the analytic turn is regrettable, these philosophers think that analysis is legitimate, but not sufficiently evidence based. For them, the pools of intuitions that analytical proposals are tested against must come from a genuine cross section of the relevant linguistic community (Alexander and Weinberg 2007). After all, they say, it is ordinary concepts that are under analysis. Experimental philosophers of *this* persuasion are easier to see as philosophers than those intent on uncovering underlying psychological processes. But the evident disunity among experimental philosophers over their purposes does not encourage confidence in a story like Knobe's, which associates philosophy in general, and experimental philosophy in particular, with an interest in underlying processes.

The third ground for questioning Knobe's argument from the history of philosophy is that it is very poor history of philosophy. First, it homogenizes into one tradition the work of preanalytic philosophers operating over millennia in very different places. Second, and more specifically, it posits a kind of smooth progress for this body of work up to the point at which, as Knobe thinks, a regrettable and decisive intellectual detour occurred—with the development of early analytic philosophy. Neither the homogenizing interpretation nor the claim that the birth of the analytic tradition is a crucial aberration is very credible. To take the homogenizing interpretation first, it seems to imply that before the start of the analytic project there was a kind of continuity in philosophy from Plato to Nietzsche. This reckons without the radical reinterpretation of Aristotelian methods of philosophy that started around the time of Descartes and continued through Hume. It also reckons without the difference made to the post-Humean understanding of nature by Kant, and through Kant, to the understanding of nature in Nietzsche. Though it may be true that there is a sustained interest from Plato to Hume to Nietzsche in empirical questions, the existence of that very low common denominator does not mean that there is a uniform, or even a relatively unitary, understanding of the empirical from ancient to preanalytic philosophy. On the contrary, there is reason to think that Aristotelian natural philosophy was much more geared to preserving prephilosophical opinion or "appearances" than either Plato before him or the post-Cartesian philosophers. Indeed, the concept of scientia, which on the surface appears to be shared by ancients and moderns, is in fact radically reinterpreted, starting with Descartes (Sorell, Rogers, and Kraye 2010). As for the supposed status of the analytic tradition as an antiempirical turning point in philosophy, this fails to take into account Descartes's insistence on a first philosophy that is prior to and conceptually independent of the concepts (extension and motion) required for physics. It also reckons without Kant's nonempirically realized conceptual apparatus for natural science.

Knobe's argument from the history of philosophy not only suggests that authentic philosophy is science. It also suggests that current physiology, chemistry, or theoretical physics is in fact philosophy, because each has emerged from the study of underlying human *and* nonhuman mechanisms that people called philosophers used to engage in. On the same grounds as Knobe argues in effect for suppression of the science/philosophy distinction, we could argue for the suppression of the philosophy/physiology distinction and the philosophy/physics distinction: both physiology and physics grew out of a much more "interdisciplinary" kind of science than we have today, the same science that is hard, according to Knobe, to disentangle from philosophy. So psychology and physics could just as well be called philosophy, and indeed might have more of a claim to count as philosophy than the analytic tradition. Certainly, this looks like a conclusion Knobe is committed to. But there is no wishing or stipulating away the conceptual change that has made *physiology*, *physics*, and *philosophy* all mean different things. And although its motivation is no different from that for saying that *philosophy* really means a broad kind of science, the claim that *physics* really means "philosophy" seems neither plausible nor interesting.

Perhaps the lesson to be drawn from the clash of experimental philosophy with the analytic tradition is quite different from Knobe's. Perhaps it is that experimental philosophy is closer in intellectual orientation to newer, more independent breakaway sciences, such as psychology or neurophysiology, than the supposedly unitary philosophy that predates the analytic tradition. Knobe cannot explain why people who currently call themselves experimental philosophers, and who are sometimes self-taught in some of the elementary experimental methodologies practiced in the social sciences, should not just get PhDs in experimental psychology and conduct mainstream research as psychologists. In other words, why insist that there is a distinctively philosophical pay-off to the experimental philosophy project when there is not supposed to be anything distinctly philosophical (as opposed to broadly scientific) for a theory of human nature or human capacities to do? With training in psychology, the people who currently work in experimental philosophy, or their successors, could tackle the traditional questions of human nature without the distractions of the analytic tradition and its alleged neglect of the big and basic questions.

But if *this* is what Knobe is recommending, it is hard to see the difference between, on the one hand, the positive program of experimental philosophy and, on the other hand, the program of naturalizing different branches of philosophy. *Naturalized* epistemology, philosophy of mind, philosophy of causation, metaethics and normative ethics, theory of reference: all of these have their proponents. The argument for naturalism does not depend crucially on an argument from the history of philosophy. It can appeal instead to the occult nature of appeals to nonnatural entities and reasons in epistemology, moral philosophy, and the rest.

The problem with some forms of the naturalizing tendency in philosophy is that they change the subject. Instead of addressing traditional philosophical problems with the help of new, including empirical, apparatus or findings, they sometimes substitute empirical for philosophical problems. For example, in Quine's seminal "Epistemology Naturalized," the task of finding skepticism-proof justification for science—roughly what Descartes attempted in the *Meditations*—gives way to the question of *how* human beings are able to arrive at empirical theories from a starting point in sensory bombardment. Again, in Churchland's version of naturalized epistemology, the task of making the justification for science skepticism-proof is abandoned, but the instrument of naturalization is neuroscience and cognitive science rather than a kind of empiricism. As in Quine, the mechanism for belief formation (or theory formation) and not the justification for beliefs formed is the focal point. Neither of these positions comes with a justification for claiming that the problem of making science skepticism-proof has been superseded, only for claiming that philosophers have not gotten far with it. So the kind of naturalism sometimes associated with Quine and Churchland, namely "replacement naturalism," is apparently more motivated by philosophical boredom—the thought that the problem is stale—than by a good argument to the conclusion that it is misconceived (Sorell 1991, 133–35).

Normative Moral Philosophy and a Second Way of Leaving the Armchair

Is there any sense in which normative moral philosophy is objectionable armchair philosophy? It is possible to answer yes to this question without thinking that armchair normative ethics is contaminated by the analytic tradition and its obsession with purely semantical issues. It is possible to object to armchair normative ethics from quite a different direction, namely the aspiration of normative ethics to guide practice and sometimes policy.

The *Stanford Encyclopedia of Philosophy* paper on experimental moral philosophy illustrates the perfectly routine dependence of some applied normative ethics on relevant empirical data:

> A philosopher considering the moral permissibility of stem cell research, for example, ought to understand the nature of these cells, gestation and embryonic development, the aims of research projects employing (or seeking to employ) stem cells, and their prospects of success. (Alfano and Loeb 2014)

In other words, philosophical work on stem cells would be deficient if it exploited only some desk-based research on the nature of these cells, but was entirely uninformed about the risks and relative success of techniques for generating organs from stem cells. These latter facts might only be well known to laboratory scientists far removed from one's armchair. In this way, incompletely informed applied ethics can, and reasonably often does, suffer from being done in the armchair.

Another way in which applied ethics can be objectionable armchair philosophy is by being naïve or utopian about the practitioners to whom it is addressed and the cultures in which they work. Philosophers teaching professional ethics to lawyers, doctors, police, bankers, intelligence agents, or military officers have no hope of influencing practice if they are unable to refer to case studies that lawyers, police, bankers, intelligence agents, or military officers would encounter in the field. Here there might be special reasons for philosophers to abjure pure thought experiment or, differently, easily conducted desk research, and inform themselves, if possible by being embedded with practitioners, what it is like in a law office, a courtroom, a police interrogation room, a police raid, or a battle. Applied ethics that is informed in this way may not reach the threshold for the kind of participant observation that is a recognized methodology of social science, but its point can be similar. It makes available to people who may have limited life experience—that is, academic moral philosophers—the phenomenology of some of the practice that normative ethics seeks to study, address, and regulate. If normative ethics is too utopian or too naïve, it loses both its audience and its opportunity for guiding the practice of those who are exposed to it.

A different set of objections is invited by normative moral philosophy when it is armchair philosophy in the sense of being too abstract or too artificial to be understood by some of the practitioners who may want its guidance. There are many illustrations of this kind of tendency in philosophical normative ethics, some prominent in the general philosophical discussion concerning whether philosophy does or doesn't rely on intuitions (Cappelen 2012).

For example, there is Judith Jarvis Thomson's argument for the permissibility of abortion after rape, with its reliance on the famous violinist example (Thomson 1971). Thomson asks the reader to imagine a woman being forced for nine months to act as life-support system, connected to a talented adult violinist in a hospital. Thomson claims that the woman is not obliged to stay connected by tubes to the violinist, because she has not consented to being a life-support system. And the violinist case is supposed to be structurally similar to pregnancy after rape. The reason Thomson uses the violinist example as the structural counterpart of pregnancy after rape is that she wishes to concede to opponents of abortion that the fetus is a fully respect-worthy human being, just as the talented violinist is. The point of her example is that even so, the woman is not obliged to act as life-support machine. She is not obliged, according to Thomson, unless she consents.

For the violinist example to work, the reader must acknowledge an analogy between the fetus and the violinist, and between pregnancy after rape and being kidnapped and tied up by tubes to a violinist who will not survive unless that arrangement is continued for nine months. The reader must also share with Thomson the intuition that acting for nine months as life-support machine is very onerous, and that there is something outrageous about being put in that position against one's will.

Thomson's defense of abortion applies not only to pregnancy after rape but to the failure of conscientious contraception. To reach the conclusion that in this rather different case abortion is also allowed, morally speaking, she imagines a world in which human beings develop from airborne people seeds that take root in houses (Thomson 1971, 59). She constructs a case in which a certain home owner has taken every possible precaution against the entrance of a people seed, and yet a people seed still enters her house and takes root, that seed then being on course to develop into a fully fledged human being. *Must* that seed be given houseroom? Another kind of seed taking root in a person's house could be removed without wrongdoing. Why not the people seed? It is not as if airborne seeds have property rights. Here again, Thomson says, leaving the people seed to develop is not obligatory but depends on the home owner's consent.

As in the violinist case, Thomson's argument depends for its plausibility on how good the analogy is between a fetus and a people seed, and between conscientious contraception and making one's house into a fortress against the incursion of people seeds. The acceptance of Thomson's defense of abortion depends on accepting the analogies. Professional philosophers and some philosophy students will not be put off by the artificiality of the examples or the science fiction of the people seeds. On the contrary, they will see immediately what the artificiality is *for*, which is to drive into prominence the connection between obligation and consent, and to neutralize the force of a rhetoric about each human being having the right to life. Instead of *denying* that a fetus is developed enough to have the dignity and right to life of a fully functioning autonomous human adult, which tends to lead to impasse in discussions with opponents of abortion, the argument exploits claims that are less contentious or uncontentious in the abstract, and then uses the examples to give them the sort of purchase on abortion that might produce a defense.

There is a connection between the persuasiveness of Thomson's argument, when it *is* persuasive, and reflective equilibrium (Brun 2014). The intuitions elicited by the examples support the substantive conclusion that permission to terminate a pregnancy depends heavily on the pregnancy being against the will of the woman in question. But the way that being against the will gets purchase is by way of examples of onerous commitments that are coerced, and failed sustained attempts at preventing something. The conclusion is supported if the intuitions elicited by the examples are to do with coercion and failed prevention, but not otherwise. And the persuasiveness of the conclusion reinforces the felt relevance of coercion and failed prevention to the question whether something is obligatory.

The fact that reflective equilibrium is reached about the importance of consent in cases of pregnancy through rape and failed but conscientious contraception does not of course mean that reflective equilibrium is reached about the permissibility of abortion in perhaps the more typical case—where pregnancy results from failure to use contraception. But the cases where Thomson is persuasive may provide leverage for the typical case.

To the extent that it diverts the argument for abortion away from its usual dead end, the armchair approach to the philosophy of abortion—the purely thought-experimental approach—is entirely defensible. But it has obvious limitations if directed at employees in abortion clinics, or doctors and nurses who have doubts about carrying out abortions, and would like to find solid moral justification for what they are doing. *This* potential audience for applied ethics is better served by philosophers who engage with case studies representative of the reasons why women present themselves at abortion clinics, the life chances of women who carry teenage pregnancies to term, social interactions in abortion clinics, and the politics and practice of pro- and anti-abortion activists.

Call this *practitioner-facing applied ethics*. I am not suggesting that this kind of applied ethics is necessarily in conflict with abstract and thought-experimental armchair applied ethics. On the contrary, the two are complementary. The audience for abstract and thought-experimental armchair applied philosophy may be other philosophers primarily, but it can include practitioner-facing applied ethicists, and it may be through its influence on practitioner-facing applied ethicists that abstract and thought-experimental applied ethics has any influence on practice at all.

The practice that applied ethics seeks to influence is not only of one kind. Business practice is irreducibly connected with self-interest, and with financial rewards, sometimes disproportionately; military practice seeks to minimize loss of life but not to avoid it. The cultures that have grown up among businesspeople and the military, respectively, are therefore morally different from that of members of religious orders who give up all material possessions, or those who are professionally devoted to saving lives. This means that business *ethics* and military *ethics* carry risks of moralism that other branches of applied ethics may not. Moralism is at one extreme of the risks posed by business and the military for applied ethics. At the opposite extreme is the risk of going native, of seeing everything as typical businesspeople and military people see things, and of waiving moral norms as if embracing *any* were to succumb to moralism.

Clearly, something between the extremes is to be aimed at. Applied ethics must be informed enough to mean something to, and command the interest of, people who have *been* there—who have taken part in military operations or who have felt the urges of ruthless business competition, or who have tried to save lives in makeshift medical facilities. So applied ethics must leave the armchair (Sorell 1998). But it must also be connected enough to systematic moral philosophy to be able to add to casuistry conducted by journalists, vicars, or politicians. Being empirically informed seems necessary to being an adequate practitioner-facing applied ethics, but it is not at all clear that the appropriate way of being empirically informed is by carrying out experiments.[8] Experimental philosophy may make sense as a response to the Thomson examples, or to philosophical discussions of the trolley problem, but it is hard to see how it contributes to good practitioner-facing applied ethics. Or, in other words, not every departure from the armchair, even in applied ethics, is a departure in the same direction.

Long before experimental philosophy came along, there were arguments in various branches of applied ethics about how empirically informed they should be (Hope 1999). It is striking that these debates had something of the impetus driving experimental philosophy: namely dissatisfaction with the dominance in bioethics of philosophers who had not left their armchairs. Similar debates have taken place in business ethics. The way that these debates have been settled is by means of a multidisciplinary and interdisciplinary approach to applied ethics. Sociology, anthropology, and even history have been applied as disciplines to ethical questions in medicine and business. Call this the *eclectic turn* in applied ethics.

If the eclectic turn is fruitful, then applied ethics will be less and less easy to regard as a subdiscipline of philosophy. Leaving aside the question of whether this is a good thing, why isn't the eclectic turn at least as intelligible a reaction to armchair philosophy as the acquisition of experimental methodologies by philosophers? In other words, why do experimental philosophers adopt the do-it-yourself approach when, except in the relatively narrow area of the appeal to intuitions within philosophy, there are common subject matters in and out of philosophy for philosophers and nonphilosophers, using the distinctive methods of many disciplines, to study cooperatively?

I venture the following explanation of the do-it-yourself approach. Experimental philosophers think that if they do not *themselves* bring empirical methods to the study of intuitions, there will be nothing, or nothing much, to link philosophy to natural science, and they believe that the link to natural science is a lifeline. The alternative to a philosophy linked methodologically to science, they think, is a moribund discipline endlessly debating semantic questions. I believe that reflection on normative ethics and its constraints discredits this way of thinking. Very little philosophical normative ethics, from the grand theories of Rawls, Dworkin, Parfit, and Scanlon, to abstract and thought-experimental applied ethics, to practitioner-facing moral casuistry, is an exercise in semantics. What keeps it alive is human life itself—individual and collective practical life, and, within collective practical life, shifting demands on role holders in social institutions. It does not have to be parasitic on the body of natural science.

Conclusion

Although the criticism of armchair applied moral philosophy that I have been mounting parts company with scientistically inspired criticism of armchair epistemology and metaethics, part of their motivation is the same. In each case, one feels dissatisfied with philosophy that is not sufficiently and appropriately empirically informed. But we can be in favor of empirically informed epistemology, metaethics, and normative ethics without being supporters of experimental philosophy. It is only when empirically informed philosophy is assimilated to science—as in certain forms of naturalism or in certain defenses of experimental philosophy—that things go badly wrong.

Acknowledgments

I am grateful to Alberto Vanzo and Johnnie Pedersen for valuable comments on an earlier draft. Work on this chapter was supported by the Arts and Humanities Research Council Xphi project, grant AH/L014998/1.

......

Notes to Chapter Fourteen

1. In a recent interview (http://philosophycommons.typepad.com/disability_and
_disadvanta/2016/06/dialogues-on-disability-shelley-tremain-interviews-joshua-knobe
.html, accessed January 19, 2017), Joshua Knobe has objected to defining experimental philosophy *in general* by its hostility to intuitions. "It is an extraordinarily diverse movement, knit together only by a commitment to conducting systematic experimental studies, and different experimental philosophy projects have very different aims." He concedes, however, that some of that movement *is* hostile to intuitions, and this is perhaps the branch of experimental philosophy whose challenge to mainstream philosophy is easiest to identify.

2. This is probably in keeping with what Maarten Boudry calls "naturalism." See his chapter in this volume.

3. In this volume, Stephen Law also associates the characteristically conceptual character of some philosophical problems and solutions with work that can be done from the armchair.

4. "A Plea for Excuses?" is J. L. Austin's most explicit piece on the method of ordinary language philosophy. There he actually seems to anticipate the language of experimental philosophy. He says that the study of excuses is "a good site for *fieldwork*" in the study of Freedom and Responsibility (Austin 1969, 183). The sources of fieldwork are the dictionary (186), the law (187), and psychology (189).

5. For the importance of ambiguity in weighing up the prima facie surprising results of experimental philosophy, see Sosa 2006, 103.

6. For example, see Williams 1978, 61 on the connection between the concept of knowledge and a God's-eye or "absolute" conception of facts or the whole set of facts.

7. But see Sosa 2006, 104 on how some of these hypotheses are more complicated than hypotheses of verbal disagreement that would come naturally to nonexperimental philosophers.

8. For a very up-to-date book of essays that seems to me to display very thoroughgoing uncertainty about the relation between empirically informed and experimental approaches, see Christen et al. 2014.

Literature Cited

Alexander, J., and J. Weinberg. 2007. "Analytic Epistemology and Experimental Philosophy." *Philosophy Compass* 2: 56–80.

Alfano, M., and D. Loeb. 2014. "Experimental Moral Philosophy." *The Stanford Encyclopedia of Philosophy* (Winter 2016 ed.), edited by Edward N. Zalta, http://plato.stanford.edu/entries/experimental-moral/.

Austin, J. L.1969. "A Plea for Excuses?" In *J. L. Austin Philosophical Papers*, edited by J. O. Urmson and G. L.Warnock, 175–204. Oxford: Oxford University Press.

Beebee, H., and N. Sabbarton-Leary, eds. 2010. *The Semantics and Metaphysics of Natural Kinds*. New York: Routledge.

Brun, G. 2014. "Reflective Equilibrium without Intuitions." *Ethical Theory and Moral Practice* 17: 237–52.

Cappelen, H. 2012. *Philosophy without Intuitions*. Oxford: Oxford University Press.

Christen, M. C. van Schaik, J. Fischer, M. Huppenbauer, and C. Tanner, eds. 2014. *Empirically Informed Ethics: Morality between Facts and Norms*. New York: Springer.

Coady, C. A. J. 1992. *Testimony*. Oxford: Oxford University Press.

Hope, T. 1999. "Empirical Medical Ethics." *Journal of Medical Ethics* 25: 219–20.

Kauppinen, A. 2007. "The Rise and Fall of Experimental Philosophy." *Philosophical Explorations* 10: 95–118.

Knobe, J. 2007. "Experimental Philosophy and Philosophical Significance." *Philosophical Explorations* 10: 119–21.

Nichols S., and J. Knobe. 2007. "Moral Responsibility and Determinism: The Cognitive Science of Folk Intuitions." *Nous* 41 (4): 663–85.

Sorell, T. 1991. *Scientism: Philosophy and the Infatuation with Science*. London: Routledge.

———. 1998. "Armchair Applied Philosophy and Business Ethics." In *Business Ethics: Perspectives on the Practice of Theory*, edited by C. Cowton and R. Crisp, 79–96. Oxford: Oxford University Press.

Sorell, T., G. A. J. Rogers, and Jill Kraye, eds. 2010. *Scientia in Early Modern Philosophy: Seventeenth-Century Thinkers on Demonstrative Knowledge from First Principles*. Dordrecht, Netherlands: Springer.

Sosa, E. 2006. "Experimental Philosophy and Philosophical Intuition." *Philosophical Studies* 132: 99–107.

Swain, A., and J. Weinberg. 2008. "The Instability of Philosophical Intuitions: Running Hot and Cold on Truetemp." *Philosophy and Phenomenological Research* 76: 138–55.

Thomson, J. J. 1971. "A Defense of Abortion." *Philosophy and Public Affairs* 1: 47–66.

Weinberg, J., S. Nichols, and S. Stich. 2001. "*Normativity* and Epistemic Intuitions." *Philosophical Topics* 29: 429–60.

Williams, B. 1978. *Descartes: The Project of Pure Enquiry*. Harmondsworth: Penguin.

15

Against Border Patrols

MARIAM THALOS

Introduction: The Ascendancy of Science

Science, physics especially, has enjoyed a long history; it has been around since the time of the ancients. This is not to say that it has always been esteemed by ordinary people, whether in its products (theories), in its personnel (individual scientists), or simply as a cultural institution. The high status science enjoys in our own era is a recent development, no more than about a century old. Jimena Canales (2015) documents the meteoric rise of physics into high repute in the early twentieth century. She argues that this ascent came on the heels of a certain confrontation with philosophy, which at the time enjoyed much greater respect. At the epicenter of the encounter was a debate that took place in Paris in April 1922, on the subject of the nature of Einstein's conception of time in the theory of relativity. This was an encounter between the then-celebrated French philosopher Henri Bergson and the then-not-all-that-famous Albert Einstein. Physics, according to Canales, enjoys its current ascendant status and muscular image at the expense of philosophy—one might even say the "feminization" of philosophy in the eyes of the larger world as an enterprise concerned more with subjectivities of experience (for instance the experience of time) than with objective and

283

immutable realities.[1] Canales describes the "backstory of the rise of science" as one achieved at the price of antagonistic relations with other disciplines, philosophy most especially; and she traces the contemporary lack of communication between scientists and humanists back to the interactions surrounding this episode in the early twentieth century:

> The scientist's views on time came to dominate most learned discussions on the topic, keeping in abeyance not only Bergson's but many . . . artistic and literary approaches, relegating them to a position of secondary, auxiliary importance. For many, Bergson's defeat represented a victory of "rationality" against "intuition." It marked a moment when intellectuals were no longer able to keep up with revolutions in science due to its increasing complexity. For that reason, they should stay out of it. (Canales 2015, 6)

To put it bluntly, Canales's (2015, esp. chap. 2) documentary evidence points to the fact that Einstein, in the follow-up to this episode, intentionally and unfairly depicted Bergson, who enjoyed a very wide readership, especially among women, as failing to understand the fundamental concepts of his theory of relativity. Einstein depicted the philosopher as concerned only with scrutinizing, defending, and employing folk concepts. He depicted philosophy, not so much as mistaken and therefore to be set right by the newest discoveries, but rather as simply and irredeemably disengaged from the realities of the physical world. "The time of the philosophers does not exist," Einstein said—not that the philosophers have made mistakes in their conceptualizations of and/or attributions to it. He positioned philosophy as ultimately incapable of the intellectual innovations required to grasp objective realities. Thus, the self-conception of so-called analytic philosophy of the mid- and late twentieth century—which persists in many philosophy departments to this day—as a discipline concerned exclusively with concepts and their application by the common folk is no accident: it seems that we have Einstein to thank, at least in part, for that self-conception.

Along with suspicion among scientists in regards to philosophy, Canales also traces the roots of logical positivism—a movement internal to philosophy considered by some in the home discipline as itself amounting to a form of scientism—to the aftermaths of the encounter in Paris between the scientist and the philosopher. Einstein was not alone in attempting to lop off some branches of philosophy from the rest of the tree of knowledge (though he was perhaps more thorough than the preeminent luminaries of logical positivism). There has been for some decades now some discussion of the extent to which philosophy has overcome the influences of logical positivism in its own ranks. Philosophy's recovery at this date can be measured in the numbers—not inconsiderable—of philosophers nowadays who take themselves, under the banner of naturalism (cf. Spurrett 2008), to be participating in an ecumenical effort to make sense of the world as a whole, including but not confined to the human experience in it, rather than as "servicing" other disciplines through analysis of concepts construed as more-or-less arbitrary human constructions.

Since the Bergson-Einstein clash, philosophers have appeared to physicists to be easy, open targets. Several prominent physicists today are reenacting this self-same stratagem from Einstein's playbook.[2] Like Einstein, they depict science as concerned with objective topics, whereas philosophy (according to their folk tales) has its gaze turned inward, as it devolves into frivolous and parochial questions. With this false view of philosophy in their sights, they propose that philosophy should be displaced by science. And some of them (prominently represented by Stephen Hawking) actually believe they're beating a dead horse, or anyway a moribund one. Some philosophers, to be sure, are prepared to bury philosophy as an aprioristic discipline, joining the ranks of those who pursue philosophical questions with empirical methodologies. May they prosper. But other philosophers, in their turn, heap condemnation upon the scientism that prompts intrusions into philosophy by scientists poorly placed to pass judgment—intrusions, according to them, where science does not belong (cf. Haack 2009).

It is my own conviction that science and philosophy are entirely too closely intertwined ever to be divorced, albeit not so closely tied as to be completely indistinguishable. Still, it would prove impoverishing, if not entirely fatal to both, should we attempt separating them forcibly. This does not mean they do not enjoy distinctive and identifying features. The possession of distinctive and identifying features, whatever that might come to, does not mean that philosophers can always be assured of being able to investigate philosophical subjects without wading into waters that physicists have navigated; and similarly in reverse: no physicists can be assured that the territories they investigate are proprietary to their discipline. What the intertwining does mean is that if we try to lop off those parts of philosophy that are heavily intertwined with the sciences, they would have to be reinvented (and badly at that) elsewhere. Where, after all, would we find a home for that discipline that critically examines (rather than superficially codifies) practices and methodologies—

principles that belong squarely in the space of epistemology? Physicists who declare that physics has the appropriate tools to study certain topics formerly thought the preserve of philosophy (for example, scientific methodology) end up making epistemological claims—they end up doing philosophy, and doing it badly.

Scientism, whatever its central features (and the subject, as this volume demonstrates, is controversial), involves poor relations between the sciences and other disciplines. If only for this, it must be combatted. But what is it to combat scientism? Many of those who wield the scientism bludgeon against others insist on better self-policing (on each side) of the boundaries between disciplines. Separate but Equal might be their slogan. Unfortunately, these users of the label *scientism*, such as, for instance (and very prominently), Susan Haack (2009)—are not in the business of illuminating for their audiences where precisely those boundaries lie. The very idea of boundaries is something that their philosophies might not even admit of, and quite typically they regard concern over the location of boundaries with suspicion—as a sign of unhealthy preoccupation with science (Haack 2009 again). So how can it make sense to insist that certain lines not be crossed when their location, and even the very existence of lines, is in question?

Now, I am myself a believer in boundaries. But I am also a believer in travel. Still, I commend mindfulness when crossing boundaries. (I will by the end of the chapter have revealed my commitments regarding boundaries between disciplines.) But while I think it's healthy to be aware of the existence of boundaries, I do not favor policies of securing the boundaries. I think some boundary transgressions are healthy and advance knowledge—*cross-fertilization* is a better word for these crossings than *transgression*. To be sure, there are bids at hostile takeover that must be forcibly repulsed. They are *reductions*, and their exercise must be carefully policed. I am against certain hostile takeovers; but my reasons are neither because they are reductions nor because they are "scientistic," but only because they rest on demonstrably bad arguments. (Some takeovers rest on unexceptional argument, and in that case the takeover must be allowed.) However, no amount of border patrolling can reveal the flaws in a bad argument for hostile takeover. Fallacies must be exposed for fallacies in the usual way—through philosophical examination of the takeover arguments. (Philosophy is exceptionally good at exposing weaknesses of this sort.) So the "foul" I shall be calling is not the foul of scientism, or even the foul of reductionism (since some reductions are perfectly warranted). Rather, I shall call foul on hostile takeover bids that rest on fallacies.

These perhaps require no special name, since "fallacy" is severe enough—indeed the most severe foul that a philosopher can call. Still, a moniker can help us notice similarities among such fouls: the term *reductionism* has already been recruited for this purpose. I shall demonstrate the moves of exposing the fallacy of reductionism by looking at the curious case of sociobiology.

If we are after better relations between the sciences and the rest of the intellectual world—and indeed we should be—we should endorse clear-headed appreciation of the virtues of scientific inquiry, as well as how science (the good stuff) is further improved by the well-functioning of other disciplines with which it has ties. I hope to provide some appreciation of that here. But I reject "separate but equal"; it motivates territoriality and bids at hostile territory grabs. At the very least, it sets the stage for border skirmishes on a grand scale. And these are simply counterproductive, because border crossing can be good for all disciplines concerned (Thalos 2013a).

Science as Public Reason

Science is neither a philosophy nor a belief system. It is a combination of mental operations that has become increasingly the habit of educated peoples, a culture of illuminations hit upon by a fortunate turn of history that yielded the most effective way of learning about the real world ever conceived.

E. O. Wilson (*Consilience*, 45)

Whatever science is, it commands respect—and rightly so. Science represents hard-won knowledge. That's why there are so many pretenders (known to their detractors as *pseudosciences*). And that's also why the word of high-ranking scientists carries weight even when they speak on topics outside their areas of expertise.

If we conceive of science the way that E. O. Wilson's statement encourages—or, in more academic terms, as *theoretical reasoning*, a conception very close to that put forward by Paul Hoyningen-Huene (2013)—we should find no fault in the idea that science is to be called on in treatment of all questions of how the world happens to be. If we adopt this conception, there can be no question of overenthusiasm for science, no question of uncritical deference to it, no concern for policing its borders, and no worries over its unwillingness to acknowledge its limitations (such concerns as feature in a highly influential recent article by Susan Haack [2009]). For, according to this conception, science simply has no boundaries. If, on the other hand, we should define science more narrowly (ghettoize it, if you will), as that which is typified in the activities or products of physicists, chemists, and maybe also biologists — the so-called *natural scientists*—then obviously there should be great concern for ensuring that the goblin does not emerge uncritically from its ghettos to infiltrate where it has no business. Then, and only then, should we be concerned about scientism. Obviously, then, a great deal hangs on how we conceive of science.

The fact that so much depends on the choice of conception is reason enough to engage with old problems about the borders of the sciences — reason enough for those who are concerned about scientism to concern themselves about lines of demarcation. For there can be no serious worry about illicit crossing of boundaries, if the borders are themselves fluid or lacking stability in the academic imaginary. It is a simple philosophical mistake, for those who worry about science overstepping its boundaries, to proclaim that identifying those boundaries is of no moment.

My own vote is in favor of a conception of science roughly like that of E. O. Wilson's. Hence, I cannot acknowledge the possibility of overenthusiasm for or unwarranted deference to science. All enthusiasm for science is warranted, so long as we recognize that science is inherently fallible — that it's not mathematics — and moreover that science cannot guarantee that technology will eventually cure all our ills. Healthy enthusiasm for science can follow from a true appreciation of what science is and what it's not: as divine as it might be, science is not another name for Almighty God.

This does not, however, mean that I agree with certain high-profile scientists when they pronounce the premature death and displacement of philosophy. They are committing simple-minded intellectual errors, to say nothing of the gross discourtesy and grandstanding on the part of a few. My aim here is not to take these confrontations case by case. Rather, my aim is, first, to illuminate the true threats to our common knowledge enterprise, via a case study of sociobiology's hostile bid for takeover vis-à-vis the subject of explaining human behavior. My objective is to illuminate a class of mistake which threatens the joint enterprise by shrinking the space of that which deserves illumination. A diagnosis of "reductionism" (not mere reduction, but reduction via bad argumentation) will be in order — it is a threat that has earned a special label. My aim ultimately is to contrast this threat (already bearing its special label) with the simple-minded challenges that we are being encouraged to label *scientism* and diagnose as some sort of border transgression. The point shall be that reductionism as a threat calls forth vigilance, but borders need no special policing. Moreover, calling out a foul on reductionisms is not a form of border patrol. To make this case, we have to be clear what we mean by *science*, and have a sense of what we mean by *borders* even if we cannot say where exactly they fall. So let's begin there.

I take science to be a public, not a private, knowledge enterprise. This is a loose characterization of science, and so not an especially controversial one. What it is meant to do is to correct a certain hyperindividualism (present, for example, in the Wilson quote at the head of this section). The contention is that science is of necessity something that is open to (and indeed has been subjected to) public scrutiny; that science is a culture of generating and curating a body of ideas/concepts/theories/evidence or what-have-you that have been so scrutinized. A person's private thoughts, however well conceived, don't qualify as science simply for being good thoughts. This is not a place for advancing arguments for this contention, but I will mention one reason for thinking in these more public-oriented terms: while individuals contribute many different kinds of things to science as individuals, what they cannot do is provide quality control. QC is a systemic undertaking provided to science by scientists as a group. And science is nothing at all if not something subjected to stringent quality controls.

Answering the question of what science is has indeed been construed as a sign of an unhealthy preoccupation or obsession with science—as a sign of scientism (Haack 2009 again). Even though the topic launched the whole discipline of philosophy of science in the twentieth century, it is still controversial, and generates renewed interest now and again. And indeed, there's no way around grappling with this question in an honest way, in a world in which the amassed bodies of work published in highly ranked journals (never mind the effluvia outside these venues) outrun any single person's ability to master the material in question. We live in a world in which no single individual can certify the value of everything that relentlessly assaults us by way of claims to truth. And indeed, we've always lived in such a world, but the magnitude of this problem has been growing exponentially for several hundred years. We feel the problem most acutely when we consider the education of the most intellectually vulnerable among us—the children. We ask ourselves: to what do we want these young minds exposed by way of authoritative ideas? And the question of how to define science then becomes quite urgent.

Scientism and Boundaries

Scientism has been multiply defined:

- a term used to refer to belief in the universal applicability of the scientific method and approach, and the view that empirical science constitutes the most authoritative worldview or most valuable part of human learning to the exclusion of other viewpoints (Sorell 1994)
- "the view that the characteristic inductive methods of the natural sciences are the only source of genuine factual knowledge and, in particular, that they alone can yield true knowledge about man and society" (*New Fontana Dictionary of Modern Thought*)
- the dogmatic endorsement of scientific methodology and the reduction of all knowledge to only that which is measurable (Outhwaite [1988] 2009, 20ff.)
- "pejorative term for the belief that the methods of natural science, or the categories and things recognized in natural science, form the only proper elements in any philosophical or other inquiry" (Blackburn 2005, 344)
- indicates the improper usage of science or scientific claims. This usage applies equally in contexts where science might not apply, such as when the topic is perceived to be beyond the scope of scientific inquiry, and in contexts where there is insufficient empirical evidence to justify a scientific conclusion. It includes an excessive deference to claims made by scientists or an uncritical eagerness to accept any result described as scientific. (Haack 2007; 2009)
- "science, and only science, describes the world as it is in itself, independent of perspective" (Putnam 1992, p x)

We can take each of these as a suggestion for how to use the expression. Each suggestion characterizes a pejorative usage, a term of abuse, but without offering a usable criterion for serious diagnosis. These (multiple) definitions can be put to diagnostic purposes only if one already knows what *science* denotes. Some authors of other chapters in this volume would like to use the term *scientism* more constructively, less vaguely, without the shaming, as a badge of honor.

A preponderance of usages of the term as a term of abuse—if not merely as a waggle of the finger or a clicking of the tongue—function to mark the target of criticism as having transgressed a boundary. And all too often it is taken for granted, without argument, that boundaries are clearly established. This is not helpful. And all too often the term *scientism* exaggerates the distance and heightens the antagonism between the parties without clarifying the underlying sources of antagonism (which are often personal grudges or jealousies)—thus generating more heat than light. But there is a common enemy worth patrolling against, as I will now explain. And it has all the surface features of a boundary dispute, without actually being such a dispute. Its name is reductionism.

The Common Enemy: Reductionism

I propose in this section to examine at some length an example of a purported boundary skirmish. I will pronounce a diagnosis of "reductionism" upon it, owing to faulty reductionist reasoning ("nothing-but"ery). The prognosis is good relations between the sciences of biology, sociology, and philosophy, among others, on condition that the nothing-but-ery is rejected.

Ever since E. O. Wilson published his seminal work *Sociobiology* (1975), research on human ethology (of quite variable scientific quality) has been conducted in departments of ethology, anthropology, biology, and now evolutionary psychology. And throughout that time, this research has been protested by both academics and laypeople, and even by highly respected biologists (cf. Gould and Lewontin 1979) as bad science as well as bad social policy. The critics have complained that sociobiologists following Wilson's program are committed to (among other things) a form of genetic determinism and an excessively enthusiastic adaptationism (bad biology), and that they tend to be dismissive of the contributions of learning and culture (bad science generally), all of which have knock-on effects of harshening social conditions on people who are still socially disadvantaged (bad policy) (cf. Downes 2014; Driscoll 2013).

The fundamental premise of the sociobiological program is that human behavior is subject to explanation via an adaptationist model: behavior B prevails in a variety of human behavioral contexts today because it conferred a higher fitness than the (behavioral) alternatives to it at a time when natural selection was acting on various aspects and "internal mechanisms" of the human brain. This highly controversial model for explaining human behavior has proliferated in recent decades, flowering into such fields as *evolutionary psychology*, *human ethology*, and *evolutionary* anthropology, among other labels.

The founding argument of the research program is roughly this one, advanced by Wilson in a number of publications (see especially 1978):

1. Biology explains animal behavior, utilizing principles of natural selection.
2. Human beings are biological, and therefore subject to biological principles like natural selection.
3. The human brain, like any other organ, is a device for survival and reproduction.

    ... Therefore, biology explains human behavior too, utilizing principles of natural selection.

While here is not the place to mount a defense of my view on a range of sociobiological explanation, I will venture a brief summary. My view is that the sociobiological model of explaining behavior is a fine model, but one model among many eligible models, many of them framed entirely in nonbiological terms. But so long as standards of evidence for applying it are respected, it is eligible to compete alongside other eligible models. The model is ineligible, however, if the (quite high) evidential standards for proposals purporting evolutionary adaptations cannot be met. Meeting these standards is important to ruling out competing models of explanation (including other evolutionary models). At the heart of many critic complaints of sociobiological models is the concern that many sociobiologists neglect their due diligence vis-à-vis these standards of evidence. And while many critics of sociobiology would simply like to reject all sociobiological proposals, I have no such agenda. I simply insist that sociobiologists make their case, like all other biologists, one argument for one behavior at a time. And this means that the general argument above—that would with one stroke sweep away all competing models of explaining behavior—is flawed. As a philosopher, I shall undertake the task of exposing the flaws, naming them, and making a motion to dismiss the argument from the legacy of knowledge to pass down to our children.

Before exposing this argument's decisive flaw, I draw your attention to the second premise of the argument, which states that human beings are biological. It is most definitely a true statement, so the flaw in the argument is not to be located here. This true statement, however, often does double duty: it is sometimes understood as saying that human beings are biological—and nothing more. This is the nothing-but-ery that qualifies the stance of this argument as reductionistic. It is the reductionism for which we shall impugn the argument, not the simple fact that it declares human beings to be biological—among other things.

Here now is my argument to expose the reductionism in this argument—the flawed attempt to dismiss competitors to sociobiology with one sweep. If we accept the argument above as valid, we must also accept as valid a second argument with exactly parallel logical structure but slightly different content words (differences italicized):

1. *Physics* explains the behavior of *physical bodies*, utilizing *physical principles*.
2. Human beings are *physical* bodies, and therefore subject to *physical principles*.
3. The human brain, like any other organ, is a *physical body*.
    . . . Therefore, *physics* explains human behavior, utilizing *physical principles*.

If we accept the conclusion of this parallel argument, we are explicitly denying the need for biological theory—indeed for evolutionary theory—as independently valuable in the enterprise of scientific explanation. Whence biology, as such, is made irrelevant. And sociobiology swept away with it. The logic of the reductionistic argument sweeps the board of all but physics.

Thus, if we accept the original argument as valid, we are obliged to conclude that it is a self-neutralizing threat. Better to deny the validity of the original argument: if biology is not to be made an irrelevant theory, then it cannot be thought to threaten other, otherwise well-supported, theories of human behavior, *simply on the grounds that human beings are biological too*. When the argument is revealed as reductionistic, the would-be reduction can be rejected purely on that basis.

The point is *not* that all reductions have to be banned, but that reductions, when they are successful, are acceptable *not* on the strength of reductionistic arguments (like the one we have just examined) but instead on the strength of the evidence they present for the proffered model of scientific explanation of the specific phenomena under scrutiny. We must police to ensure that when a reduction (or indeed any explanation) is proclaimed, the evidence for it is in good order; it is just another day in the academy. No extraordinary measures are required. And certainly no special measures to ensure that biologists remain in their corner, and everyone else do likewise.

The End of Acrimony? Let's Return to an Old Theory of Science

C. P. Snow, who coined the term "two cultures" to describe the divide in our contemporary era between the sciences and the humanities, bemoaned that divide, and held that it kept humanity from solving many of its problems. If Canales's research sheds light on anything, it sheds light on the fact that the divide is entirely an accident of history. Things could have gone differently. Had there been nothing to be gained by appearing to be at odds with each other, Einstein and Bergson might well have come away with an understanding of each other as working in complementary ways rather than at odds. And even if one of them (Bergson, say) was in error, the other could have acted as a corrective. The dispute need never have gone to a place where one party was accused of chasing after unicorns in a chariot of fire. The rift was not inevitable—and certainly not required by the logic of each of the disciplines in question.

I have thus far refrained from advancing here any controversial theories of science, in favor of something quite neutral. But it will not serve us to proceed in this vein if we are to heal the rift between the sciences and the humanities—if we are to correct misconceptions of each by the other. Those disciplines routinely accepted as natural sciences (physics, chemistry, biology, most prominently) do not share a common core, either in methodology or in any other way. What's more, there are counterexamples in their combined bodies' narratives to almost every philosophical account of a science. Except the one I shall advance. My goal in outlining this theory here (I've defended it to a much greater extent in Thalos 2013a) is nowise to advance any sort of demarcation criterion between the sciences and everything else—which is in any case a hugely ambitious task. The goal instead is to illuminate the commonalities between sciences and many nonscientific enterprises, including philosophy. The aim is to illuminate the notion of public knowledge, and so display what they have in common rather than what divides them. Why they are entwined and cannot be prised apart. Why their kinship allows them to function well when they learn judiciously from one another. And I don't claim any originality in this theory of science. Indeed, it is the oldest theory of science that there is. It was framed by the great ancestor of all the natural sciences—Aristotle.

It is undeniable that Aristotle made the largest and earliest contributions to advancement of the very conception of science. In my estimation, his insights are unsurpassed. He began with the space of reasoning. Figure 15.1 depicts the

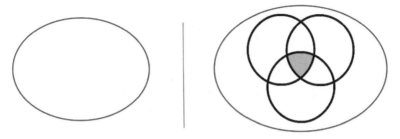

Figure 15.1. The Aristotelian space of reasoning forms.

Aristotelian picture of the space of reasoning, and reasoning practices. I attribute the first articulation of a gulf between theoretical and practical reasoning back to Aristotle, who distinguished strongly between common sense (where that also included common or "folk" theories of everything) and disciplinary thinking, which includes distinctive disciplinary modes of reasoning (represented by the overlapping circles in figure 15.1). These latter are discipline-specific, refined forms of reasoning—*taught* forms that we learn to employ when we step into the role of apprentice in a scientific practice. This is a view of the space of reasoning norms that I very much support.

What I like best about this picture is the line. It represents the difference between distinct reasoning enterprises, dividing common sense (practical reasoning) from pure theoretical reasoning. Theoretical reasoning is a pure enterprise—an enterprise with a single goal: truth. *Pure* in this context means simply unmixed with other concerns. Pure reasoning is strategically simple—there is one goal only, and no other objectives are balanced against it. In pure reasoning, all the resources of inquiry are put in service of the single goal. In theoretical reasoning, one simply pursues truth; one does not work out what to do about it, or balance the value of this pursuit against the value of other pursuits. Truth is the only objective of theoretical reasoning. Of course, we are talking about the ideal enterprise of reasoning; we are not talking about reasoning as performed in real time by real individuals. The latter must be a subject for another occasion.

By contrast, because practical reasoning is allowed, even required, to comprehend many goals, it is a mixed enterprise, in which one of the primary tasks is to balance objectives one against another in a context-sensitive way—which is quite different from weighing evidence as to truth (Thalos, in progress). Some philosophers hold that truth is just another good that can be balanced against others. Others maintain that truth is not even a legiti-

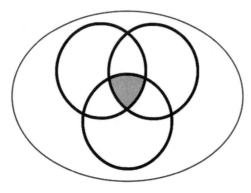

Figure 15.2. A contemporary conception of the space of reasoning forms.

mate human aim—humans aim at advantage through and through. (Some pragmatists can be read this way.) That's the subject for another occasion as well (but cf. Thalos 2013b). Long story short: for the Aristotelian, truth is an unconditional imperative; it's not a good in the sense of being susceptible to being enjoyed and balanced against other goods. This is not an implausible idea: truth is not something that is directly enjoyed and experienced by individuals. Specific truths might be directly experienced, but not truth as that which they have in common and to which knowledge claims answer. Truth, in the singular, has a dignity and not a utility. In the economy of theoretical reasoning, utility has no place.

Contrast this distinctly Aristotelian view with two contemporary views on reasoning. First is the view that reasoning is undifferentiated, with no philosophically significant lines of demarcation. Pragmatism is a species of such a view. In pragmatism, every goal is balanced and traded against every other because no aim is unconditional. Utility is the very guide of life.

A second contrasting contemporary view (fig. 15.2) depicts reasoning as indeed differentiated, but the differentiated spaces are all overlapping, and in addition there are spaces also for "all-things-considered" reasoning— reasoning that transcends and crosscuts all the specific forms, where different considerations can be balanced against each other and where tradeoffs between different concerns are adjudicated. While this picture explicitly depicts a lot of differentiation in reasoning, and boundaries between different forms of reasoning, it doesn't depict any kind of serious gulf—any lines of the sort we see in the Aristotelian picture of figure 15.1, between practical and theoretical reasoning. The point now—and this is a point with which the

Aristotelian most emphatically agrees—is that differentiation does not as such constitute a barrier to combination. The barrier or "gulf" between practical and theoretical denotes something else.

So why is there a barrier in the Aristotelian space of reasoning, whereas the contemporary picture of figure 15.2 does not even acknowledge the space marked off in the Aristotelian picture as distinct? What does Aristotle's line denote? Answering this question will take us back to where we began—back to the question of what science consists in.

Practical Reason

Aristotle, as we have already observed, distinguished strongly between knowledge and common sense. This is for a good reason. It is because he recognized the importance of reasoning as such, and the relevant discipline-relevant forms of it especially, in the securing of knowledge. For Aristotle, the reasoning basis of knowledge is deep. He held that intellectual inquiry—inquiry for the sake of truth—falls into natural categories that today we refer to as disciplines, differentiated by proprietary standards of inquiry (standards of reasoning broadly conceived). Aristotle insisted that standards of inquiry differ by discipline. For example, the standards of proof and evidence in logic and mathematics should not be applied in equal measure to the study of nature, or to politics. Three points will help us understand how Aristotle's idea is different from those that prevail today.

First, much of what counts as knowledge possession to a modern philosopher's sensibilities would not even amount to knowledge for Aristotle, since its reasoning basis is too shallow. The ability to correctly recite the price of tea in China, as a one-off event, would not count as knowledge. Second, and more important, a reasoning-oriented view of knowledge would resist the modern philosophical sensibilities that speak of having knowledge of isolated facts or propositions (*epistemically unconnected* is the technical term; cf. Hoyningen-Huene 2013). Finally, a reasoning-oriented view would center on differences between individuals as bearers of knowledge, on the one hand, and communities, on the other. In recent times we have come to appreciate that "quality control" over a knowledge repository is a community undertaking (Longino 1990, 2002; Kitcher 2001). Thus individuals, to possess knowledge, must be appropriately connected to a corresponding repository's safety nets for knowledge—to the bases of evidence (data, experiments, journal articles, etc.) as well as to networks of argumentation in support of a variety of (competing) ideas and theories. Knowledge—anything resembling knowledge of a complex world by limited knowers—will very likely bear these systemic qualities.

None of these points apply when we are considering instead the practical space of reasoning rather than the theoretical. Practical reasoning aims at action. And action is quite frequently, albeit not exclusively, something that has to be performed by individuals for personal ends (as contrasted with the aim of truth). Moreover, many beliefs are acquired simply for the sake of action—not for the sake of truth. The interests in those cases are not contemplative. These are the sorts of beliefs Aristotle refers to as common sense. They are "common" because they do not comply with specialized (discipline-specific) standards of inquiry—and are not obliged to do so. In fact, they do not comply—neither are they obliged to—with *any* standards. They are measured by the extent to which they serve their hosts, not the extent to which they adhere to standards that serve the goal of truth. Common sense answers only to its users.

And so now we have an account of the line of figure 15.1. It divides between knowledge, on one side, and other intelligent activity, on the other. Common sense is on the side of "other intelligent activity," because common sense enjoys no standards of inquiry. It is not a discipline. It is an activity of practical, as contrasted with scientific or disciplinary, life. It is not contemplative. Knowledge is relatively rare, at least in certain epochs, in that it is the product of science. It is commonplace only conditionally, and only to the extent that scientific knowledge is disseminated well and freely. If scientific inquiry had never been conducted, or if it were secreted in the temple of a holy priesthood, then there would be no knowledge spread across the face of the globe: the planet would be covered only in abject common sense, or in beliefs that have not been subjected to the high standards of evidence.

Aristotle's theory of science and his conception of knowledge are one and the same conception. I propose to use the label *science* for all of what Aristotle would have called knowledge. Hence it applies to anything that accepts the authority of a coherent set of standards of inquiry that aim at truth.[3] Coherent standards tend to be focused on special problems or topics; they comprise intellectual disciplines. Any enterprise without coherent standards falls on the side of "other intellectual activity."

Philosophy is a discipline with coherent standards of inquiry; it also aims at truth. Thus, philosophy is a science. This is not to obliterate the differences between philosophy and (say) physics. It is instead to uphold them. And to uphold them in a way that respects the separate dignities they represent. It is also to uphold the dignity of the pursuit of truth, and to distinguish it from activity that does not serve that noble goal. (It is noble insofar as its pursuit is orthogonal to pursuit of that which serves oneself.) However, this is not to denigrate intellectual activity that serves other masters: other masters must indeed be served—we aim at health, prosperity, and happiness. They are worthy goals. May the efforts aiming to advance them also prosper. But the enterprises that aim at these ends cannot, insofar as they do so, be called science. They are something worthy, but they are something else. Science is indeed a term of commendation; but it is not the only one. What, then, of scientism?

Science without Boundary Police

The account of science just laid out—meant as a revival of an old theory of Aristotle's—does speak of boundaries. They are boundaries between rather than as a cordon around the sciences. Crossing such boundaries, if done right, can serve the sciences on both sides—such as when a discovery in mathematics renders service elsewhere and when a discovery in philosophy can render service elsewhere. It is appropriate to the dignity of knowledge that it should be open—to both contributions and to appropriations—from those who are not officially initiated into the relevant specialisms. We have spoken also of enemies of the sciences—reductionisms that would obliterate entire disciplines with bad arguments. Of these everyone must be mindful, since bad arguments are not welcome anywhere. We have also spoken of common sense, as that which is not-science but which nonetheless might be useful in serving ends other than truth. When it is justified, it is justified purely in relation to those (nonknowledge) ends and by its (good) service to them.

Imagine now a hypothetical scenario: someone, perhaps in imitation of Einstein, seeking to serve his reputation at the expense of a discipline (like philosophy), and to claim his fifteen minutes in the spotlight, makes scornful remarks to the effect that philosophy is not science and hence lacks merit. Is this an intellectual error worthy of a special label, say, *scientism*? Well, it is certainly an intellectual error: the statement is flatly false because philosophy is a science, in the broad sense we have been courting. So yes, there is an intellectual mistake in this. But in a different regard, there is a sideways truth—a truth that is misconstrued: philosophy is not like physics insofar as its standards of inquiry are different. That fact, to which all will happily assent, should not be minimized. But now how about the derogation? Statements like this one can be offensive, dismissive, and condescending. And of course we can innovate a concept like scientism to fit the crime as so described. I prefer to call out the bad (meretricious) behavior, name its author as self-serving, and point out the clear fallacy. In my estimation, falsehood is a harsh enough foul. And this crime deserves no special name to raise it to heights of special scrutiny—better it should be called a flat fallacy, on all fours with the error of an Earth-centered solar system.

Thus, I hold no brief for *scientism* construed as a shaming term for border-crossers and disparagers of others' disciplines. I can think of no good service that the border patrols it speaks of can be put to. I cannot think how a regulative theory of science, one that aims to preserve its dignity, illuminate relations among the sciences, and throw light on their distinctiveness from common sense, can be improved by border patrols. The boundaries between sciences require no policing. Indeed, we are better served by scientists acting the part of friendly kibitzers, helping each other and learning from each other's findings and methods. We are better served by healthy exchanges over the fence (and under it too) than by borders to keep out intruders. Thus, my vote is to deny a mandate to would-be border patrols.

······

Notes to Chapter Fifteen

1. Canales (2015, 479) argues that Bergson himself experienced a feminization, due at least in part to his popularity among women readers and in part to the fact that Einstein never acknowledged any merits in his views.

2. The most egregious is perhaps Lawrence Krauss, who, in reaction to David Albert's negative view of his book *A Universe from Nothing*, called Albert—who incidentally holds a PhD in theoretical physics—a "moronic philosopher" in a self-servingly opinionated interview in the *Atlantic* (April 23, 2012). Wearing his ignorance of philosophy as a badge of honor, he there (falsely) proclaims that philosophy as a discipline is without merit, taking credit for work done by practitioners of other disciplines.

3. I don't say "comply," because accepting standards and actually complying with them (the difference between trying and succeeding) should not divide between science and nonscience.

Literature Cited

Blackburn, Simon. 2005. "Scientism." In *The Oxford Dictionary of Philosophy*, 344. Oxford: Oxford University Press.

Canales, Jimena. 2015. *The Physicist and the Philosopher: Einstein, Bergson, and the Debate That Changed Our Understanding of Time*. Princeton, NJ: University Press.

Downes, Stephen M. 2014. "Evolutionary Psychology." In *The Stanford Encyclopedia of Philosophy* (Summer 2014 ed.), edited by Edward N. Zalta, http://plato.stanford.edu /archives/sum2014/entries/evolutionary-psychology/.

Driscoll, Catherine. 2013. "Sociobiology." In *The Stanford Encyclopedia of Philosophy* (Winter 2013 ed.), edited by Edward N. Zalta, http://plato.stanford.edu/archives /win2013/entries/sociobiology/.

Gould, S. J., and R. Lewontin. 1979. "The Spandrels of San Marco and the Panglossian Paradigm: A Critique of the Adaptationist Programme." *Proceedings of the Royal Society London, Series B: Biological Sciences* 205: 581–98.

Haack, Susan. 2007. *Defending Science—within Reason: Between Scientism and Cynicism*. Amherst, NY: Prometheus Books.

———. 2009. "Six Signs of Scientism." University of Texas at Arlington website: http:// www.uta.edu/philosophy/faculty/burgess-jackson/Haack,%20Six%20Signs%20of %20Scientism.pdf.

Hoyningen-Huene, Paul. 2013. *Systematicity: The Nature of Science*. Oxford.

Kitcher, Philip. 2001. *Science, Truth, and Democracy*. New York: Oxford University Press.

Longino, Helen E. 1990. *Science as Social Knowledge: Values and Objectivity in Scientific Inquiry*. Princeton, NJ: Princeton University Press.

———. 2002. *The Fate of Knowledge*. Princeton, NJ: Princeton University Press.

Outhwaite, William. (1988) 2009. *Habermas: Key Contemporary Thinkers*. 2nd ed. Cambridge: Polity Press.

Putnam, H. 1992. *Renewing Philosophy*. Cambridge, MA: Harvard University Press.

Sorell, Tom. 1994. *I: Philosophy and the Infatuation with Science*. London: Routledge.

Spurrett, David. 2008. "Why I Am Not an Analytic Philosopher." *South African Journal of Philosophy* 27 (2): 153–63.

Thalos, Mariam. 2013a. *Without Hierarchy: The Scale Freedom of the Universe*. New York: Oxford University Press.

———. 2013b. "Truth Deserves to Be Believed." *Philosophy* 88 (344): 179–96.

———. In progress. "The Gulf between Practical and Theoretical Reason."

Wilson, E. O. 1978. *On Human Nature*. Cambridge, MA: Harvard University Press.

———. 1998. *Consilience: The Unity of Knowledge*. New York: Alfred A. Knopf.

CONTRIBUTORS

Russell Blackford is editor-in-chief of the *Journal of Evolution and Technology* and holds a conjoint appointment in philosophy in the School of Humanities and Social Science at University of Newcastle, NSW. His many books include *Freedom of Religion and the Secular State* and *The Mystery of Moral Authority*.

Maarten Boudry is a postdoctoral philosopher of science at Ghent University. He wrote his dissertation on pseudoscience, *Here Be Dragons. Exploring the Hinterland of Science*. On the same topic, he coedited, together with Massimo Pigliucci, *Philosophy of Pseudoscience: Reconsidering the Demarcation Problem*. Other research interests include irrationality, cultural evolution, and the conflict between science and religion.

Filip Buekens is associate professor of philosophy at Tilburg University and in the Institute of Philosophy at Katholieke Universiteit Leuven. He specializes in philosophy of language and mind, social ontology, and critical thinking. Recent publications include work on obscurantism, the philosophy of coordination, and the nature of institutions.

Taner Edis is professor of physics at Truman State University. His research interests include the history and philosophy of science, especially concerning the disreputable fringes of science.

Justin Kalef is an associate teaching professor in the Department of Philosophy at Rutgers University, where he serves as Associate Undergraduate Director and as Director of Curriculum and Online Development. Before coming to Rutgers in 2011, he worked in various departments in his home province of British Columbia, Canada. His work focuses on metaethics, particularly moral relativism.

Philip Kitcher is the John Dewey Professor of Philosophy at Columbia University.

Stephen Law is reader in philosophy at Heythrop College, University of London, and head of the Centre for Inquiry UK. He was formerly junior research fellow in philosophy at the Queen's College, University of Oxford.

Thomas Nickles is the Foundation Professor of Philosophy Emeritus at the University of Nevada, Reno, and a fellow of the American Association for the Advancement of Science. His main interest is scientific and technological innovation.

Rik Peels is an assistant professor of philosophy at Vrije Universiteit Amsterdam. His research interests are the ethics of belief, scientism, ignorance, and various issues in the philosophy of religion. Recent publications include *Responsible Belief: A Theory in Ethics and Epistemology, Perspectives on Ignorance from Moral and Social Philosophy* (editor), and *The Epistemic Dimensions of Ignorance* (coeditor).

Massimo Pigliucci is the K. D. Irani Professor of Philosophy at the City College of New York. He has a background in evolutionary biology and in philosophy of science. His interests include the structure and history of evolutionary theory, the demarcation between science and pseudoscience, and the development of the so-called skeptical movement and its impact on public understanding of science. He blogs at platofootnote.org.

Alex Rosenberg is the R. Taylor Cole Professor of Philosophy at Duke University (with secondary appointments in the biology and political science departments). He has held fellowships from the National Science Foundation, the American Council of Learned Societies, the National Humanities Center, and the John Simon Guggenheim Foundation; received the Lakatos Award in the philosophy of science; and been the Phi Beta Kappa–Romanell Lecturer. He is the author of many academic books and papers, and has written two novels.

Don Ross is professor in the School of Sociology and Philosophy at University College Cork; professor of economics at the University of Cape Town; and program director for methodology at the Center for Economic Analysis of Risk, Robinson College of Business, Georgia State University. His main research areas are the experimental economics of risk and time preference, addiction, and science-driven metaphysics. He is the author or editor of many books and journal articles.

Michael Ruse is director of the Program in History and Philosophy of Science at Florida State University. He was the founding editor of the journal *Biology and Philosophy* and editor of the Cambridge Series in the Philosophy of Biology.

Tom Sorell is professor of politics and philosophy and director of the Interdisciplinary Ethics Research Group at Warwick University.

Mariam Thalos is professor of philosophy at the University of Utah. Her work crosses borders between science and philosophy, carrying good ideas from one to serve the other, and back again in case the travel has improved it for its original home.

INDEX